Smoke in Food Processing

Author
Joseph A. Maga
Professor
Department of Food Science and
Human Nutrition
Colorado State University
Fort Collins, Colorado

CRC Press
Taylor & Francis Group
Boca Raton London New York

CRC Press is an imprint of the
Taylor & Francis Group, an **informa** business

PREFACE

Smoking was one of the first forms of food processing, and through the centuries the chemistry of smoke has slowly evolved. It is now known that wood and food composition can significantly influence smoke composition, as well as the resulting textural, sensory, nutritional, antioxidative, and antimicrobial properties of the smoked food.

Aside from beneficial properties, one must also consider potential health concerns associated with certain woods and their resulting smoke.

THE AUTHOR

Joseph A. Maga, Ph.D., is Professor of Food Science and Director of the Food Research and Development Center in the Department of Food Science and Human Nutrition at Colorado State University, Fort Collins, Colorado.

Dr. Maga received his Ph.D. in Food Science from Kansas State University and his M.S. and B.S. in Dairy Manufacturing from The Pennsylvania State University. His industrial experience includes being a Group Leader in the Research and Development Group at Central Soya and a Project Leader with Borden Foods Company.

Dr. Maga's area of specialization is flavor chemistry with special emphasis on the role of food processing and storage. Current research areas include flavor formation during thermal degradation of by-products and flavor formation and retention during food extrusion.

He is a professional member of the American Dairy Science Association, American Association of Cereal Chemists, American Chemical Society, and the Institute of Food Technologists of which he is currently President of the Rocky Mountain Section and a member of the Board of Editors for the Journal of Food Science.

Dr. Maga is the author of over 150 publications including numerous extensive reviews appearing in CRC Press publications.

TABLE OF CONTENTS

Chapter 1

WOOD COMPOSITION

I. INTRODUCTION

Before one can begin to understand the various chemical and physical interactions of wood smoke with foods, one should have an initial understanding of the basic composition of wood.

Botanically, wood used as a smoke source falls into two broad categories, namely hardwoods and softwoods, which in some cases does not correlate with the actual physical hardness of wood. For example, some hardwoods are actually quite soft (aspen and cottonwood), while certain softwoods are quite hard (Douglas fir).

The two species can be distinguished by the fact that hardwoods produce woody, two-seed leaves (dicotyledenous) and their seeds are borne in a fruit structure. Thus, they are classified as angiosperms. In addition, the leaves of hardwoods are usually broad or bladelike and in most climates they are deciduous in that they shed their leaves at the end of each growing season.

On the other hand, the softwoods are classified as gymnosperms, since their seeds are borne naked. Softwoods have also been called conifers, since many produce seed cones, pollen cones, or a combination of both. This species has needlelike (pine) or scalelike (cedar) leaves that are commonly called evergreen, since with the exception of larches and bald-cypress, the leaves are retained for up to several years. Some typical American hardwood and softwood species are summarized in Table 1.*

Wood is essentially a plant tissue composed of various types of cells. In mature wood, most of these cells are dead and hollow. This structure is known as secondary xylem, with the hollow interiors of the cells known as lumens. The orientation of cells and lumens differ between hardwoods and softwoods. In softwoods, the cells, which make up to 90 to 95% of the wood volume, morphologically appear to be fibrous in nature, and thus are called fibers. In contrast, hardwoods are composed of much wider cells, which are called vessel elements. Also, fiber volume can be quite variable for hardwood, but averages approximately 50%.

From a chemical standpoint, wood cells are composed of several organic polymers with the structural key compound being cellulose. As seen in Table 2, it is a long-chain, linear sugar molecule or polysaccharide composed of glucose units. In most normal wood tissue, glucose, as a cellulose polymer, accounts for 40 to 45% of total dry weight.

Another major constituent of wood is callea hemicellulose, which serves as the matrix substance for the cellulose superstructure. It has a lower molecular weight than cellulose and may contain short side chains. The carbohydrate composition of hemicellulose can vary and is usually composed of a combination of five-carbon sugars such as xylose and arabinose and six-carbon sugars typified by glucose, mannose, and galactose.

The third major component of wood is lignin, which serves as a cementing material that assists in the binding together of wood cells. Approximately 60% of total lignin is located in the cell walls, with the remainder spread throughout the wood. Structurally, as seen in Table 2, it is a three-dimensional, highly branched, polyphenolic molecule of complex structure and high molecular weight. It should be noted that lignin generally is not reactive with water, whereas hemicellulose and cellulose are just the opposite.

Another compositional distinction, the overall importance of which has not been evaluated

* Tables appear at the end of text.

from a smoke chemistry standpoint, is the presence of heartwood vs. sapwood. Heartwood represents the central portion and sapwood the outer portion of the wood. The heartwood portion is significantly higher in organic extractives, which are primarily polyphenolic in composition, and usually the cells in heartwood are dead and have a lower moisture content than sapwood. Also, most heartwoods take on distinctive colors such as brown, yellow, orange, and red. In addition, heartwood extractives from wood species such as redwood and cedar are toxic to microorganisms and insects. The potential implication of these toxic compounds, if any, to humans relative to smoke consumption is not known.

Another wood form that deserves special attention is bark. This by-product can be used as a wood smoke source and compositionally it is significantly different than sapwood or heartwood.

Other minor compound classes that can be present include terpenes. In softwoods they can account for up to 5% of product weight, but appear to be present in hardwoods only in trace amounts. Compounds such as turpentine and resin acids are typical of this class and can certainly contribute to the organoleptic properties of wood smoke.

Aliphatic fatty acids also occur in all woods, usually in the ester form. For most woods the ester of acetic acid is the most prevalent and can be found in wood at levels up to 5%. Dibasic acids such as calcium carbonate and calcium oxalate also are present.

Proteins represent another potentially interesting component in wood. In developing wood tissue (sapwood), they can be present in significant amounts, but in heartwood, levels of usually less than 1% are present. Protein can serve as the nitrogen source for thermally induced nitrogen-containing heterocyclic flavor compounds such as pyrazines.

Also present in small amounts are aliphatic alcohols, cyclic polyhydric alcohols such as cyclitols, and sterols, aldehydes, hydrocarbons, alkaloids, and minerals. The compositional relationships of wood are summarized in Table 3.

At this point, let us start with cellulose and expand on its composition, and then discuss in further detail the other components of wood.

II. CELLULOSE

The major component of most wood species is cellulose but it was not until the 1920s that its structure was more or less completely resolved to be a linear-polymer glucan composed of anhydroglucopyranose units bound by β1,4 glycosidic linkages.[1-3] The actual number of glucose units in one intact cellulose chain has been found to range from 9000 to 15,000[4] depending on wood source and location within the sample. Two adjacent glucose units are linked by the elimination of one water molecule between their hydroxylic groups at carbon 1 and carbon 4.

It has been shown that the two terminal glucose residues in a native cellulose chain are different from each other and different from the glucose units making up the rest of the chain.[2] One end unit contains a reducing hemiacetyl group and thus has been named the reducing end group, while the other unit terminates with an extra hydroxyl group and is known as the nonreducing end group. The structures of these two groups are shown in Table 4. Native cellulose occurs in an elongated chain with all the glucose units arranged on one plane to the β-glycosidic linkage.

The molecular weight of cellulose can vary from 50,000 to around 2.5 million depending on its origin,[18] with the size of the chain molecule usually expressed as the degree of polymerization, which in turn is defined as the ratio of the molecular weight of cellulose to the molecular weight of one glucose unit. As can be seen in Table 5, the degree of polymerization can vary with wood source. In addition, the degree of polymerization decreases as wood becomes older.

The functional groups of cellulose chains are the hydroxyl groups and there are three linked to each glucose unit. These hydroxy groups not only are important to cellulose structure but also determine physical and chemical behavior. For example, these hydroxy groups can form two types of hydrogen bonds depending on where they are located in the glucose units. One type forms intramolecular linkages which are bonding between hydroxy groups of adjacent glucose units in the same cellulose molecule. The other type is intermolecular linkages occurring between hydroxy groups of adjacent cellulose units.

Native cellulose has been shown to be partially crystalline.[6] The main unit cell is composed of eight cellobiose moieties and the molecular chains pack in layers which are held together by van der Waal's forces.[5] The chains in turn are held together by intermolecular hydrogen bonds as well as intramolecular hydrogen bonds between the atoms of adjacent glucose units.[5]

Cellulose is rather difficult to isolate in its native intact state from wood because it is intimately associated with hemicellulose and lignin, and probably as a result, other polymorphic forms of cellulose called cellulose II, III, and IV have been reported.[18] Native cellulose is called cellulose I with respect to its crystalline lattice which has three axes of different lengths and one non-90° angle. However, from an experimental standpoint, one does not have to isolate cellulose from actual wood since native cotton fiber is composed almost exclusively of cellulose.

III. HEMICELLULOSE

The term hemicellulose can be considered to be generic since originally it was thought that all wood components were derived from cellulose. Also, hemicellulose does not represent one specific compound but includes a mixture of polysaccharides that can be composed of glucose, mannose, galactose, xylose, arabinose, rhamnose, 4-*O*-methylglucuronic, and galacturonic acid residues, as shown in Table 6. Based upon the number of possible reactants, one could postulate that a large number of different hemicelluloses actually exist in nature. Some of the more common wood hemicelluloses are summarized in Table 7, and as can be seen, the compounds involved, molar ratios, types of linkage, and chain length can vary significantly.

In general, hemicellulose constitutes 20 to 35% by weight of wood, has a lower molecular weight than cellulose, and can be branched as shown in Table 8. The sugar units making up hemicellulose can be subdivided into pentosans, hexoses, hexuronic acids, and deoxyhexoses. In turn, the main chain of a hemicellulose can consist of only one unit forming a homopolymer such as xylans, or if two more units are present, a heteropolymer such as a glucomannan can form. In addition, some units such as 4-*O*-methylglucuronic acid and galactose are usually found as side groups of a main chain.

Hemicellulose is closely associated with cellulose via hydrogen bonding and van der Wall's forces, and provides structural integrity to wood. In addition, it can be present in high amounts when a tree is under stress and is rather easily hydrolyzed by acid.

Woods differ not only in the proportion of total hemicellulose present but also differ in the percentages of individual hemicellulose compounds due to differences in individual carbohydrate units. This latter compositional difference is summarized in Table 9. In general, softwoods have a higher proportion of mannose than hardwoods but a lower proportion of xylose.

Hemicellulose can chemically be divided into xylans typified as arabinoxylans, glucuronoxylans, and arbinoglucuronoxylans; mannans such as glucomannans and galactomannans; glucans; galactans such as arabinogalactans, and pectins.

Xylans are homopolymer hemicelluloses composed of xylose units. In addition, hardwood xylans usually contain small amounts of rhamnose and galacturonic acid. Hardwood xylans have 4-*O*-methylglucuronic acid units at irregular intervals bound by α–1,2 glycosidic linkages. In contrast, softwoods lack acetyl groups but have arabinofuranose units linked α−1,3 by glycosidic bonding.

Mannans can be characterized as having a heteropolymeric backbone composed of mannose and glucose units. Hardwood mannans are only slightly branched while softwood mannans have acetyl and galactose residues attached. Softwoods normally contain 20 to 25% mannans.[19]

The major glucan in wood is starch, which exists as granules in both the linear amylose linked α−1,4, and the branched amylopectin fraction linked α−1,6. Another glucan is callose, which is linked β−1,3. Other unique minor occuring glucans can be found in certain woods.[19]

Galactans are water soluble and can be found at levels of from 10 to 25%. In general, galactans are highly branched with the galactans of hardwoods having a fair amount of rhamnose present.

IV. LIGNIN

Lignin is the third major component of wood, with mature wood containing between 18 and 38%,[8] as can be seen in Table 10. Chemically, lignin is a copolycondensate of the dehydrogenation products derived from the hydroxycinnamyl alcohols *p*-coumaryl-, coniferyl-, and sinapyl alcohols, whose structures are shown in Table 11. The resulting product can be classified as a phenolic-based compound containing numerous possible combinations of differently bonded hydroxy- and methoxy-substituted phenylpropane units. Some of the most common types of bonds, along with their relative amounts of hardwood and softwood are summarized in Table 12. The alcohols can be linked by numerous types of bonds including ether (phenolic ether and dialkyl ether) and carbon-carbon ($5,5^1$, $\beta\text{–}^5$, $\beta\text{–}\beta^1$, and $\alpha\text{–}\alpha^1$) bonds. Thus, the number of intermediates can be quite large. For example, Freudenberg[9] was able to isolate 20 intermediates just from the dehydrogenation of coniferyl alcohol. Certain woods may also have *p*-hydroxyphenylpropane units that can participate in lignin formation.[8] As a result, lignin is probably one of the most complex naturally occurring compounds relative to structure and heterogeneity. The branched polymers that result lack regular and ordered repeating units thus making compositional identification difficult. Also, during the isolation of lignin from wood, degradation usually occurs again complicating identification. In addition, polysaccharides usually are contaminants in isolated lignin since they form strong complexes with lignin.

Lignins in turn can be divided into two major classes,[20] based on their structural elements. One such class is guaiacyl lignin, which is usually found in softwoods and represents a polymerization product of coniferyl alcohol. Another is guaiacyl-syringyl lignin, which is found in hardwoods.

Aside from the major components discussed above, wood contains many other types of compounds that historically have been grouped together and given the generic name of wood extractives. In general, these compounds are partially soluble in polar and/or nonpolar solvents and constitute a few percent of wood grown in temperate zones but can be relatively high in woods from subtropical and tropical areas. In addition, their composition may be high in certain parts of a tree such as branch bases, roots, and areas of irritation.[21]

Extractives are important in that they can provide characteristic color and/or odor to certain woods. In addition, certain extractives are toxic or act as a deterrent to certain bacteria, fungi, and termites.

During the aging of harvested wood the extractives composition can change significantly, resulting in the loss or degradation of unsaturated compounds, fats, and fatty acids. Thus, green and aged woods will have different precursors available for the formation of compounds during smoking.

Also, as would be expected and as shown in Table 13, the extractives composition varies among wood species. Also as seen in Table 13, the solvent of choice will significantly influence the proportion of extractives removed. In addition, climate and season can influence extractives composition.[21]

Another generic term associated with wood extractives is resin, which really is not representative of certain specific compounds but represents a physical condition. Resin can be composed of terpenes, lignans, stilbenes, flavonoids, and other aromatics which mutually inhibit crystallization and it is usually more predominant in softwoods than hardwoods.

At this point some of the major compound classes that go to make up the extractives and other miscellaneous compound classes found in wood will be discussed.

V. MISCELLANEOUS COMPOUND CLASSES

A. Volatile Oils

Distinctive differences in the odors of various wood species are due to the volatile or essential oil composition of each specific wood. Dependent upon wood species, the actual compounds responsible for odor may be terpenes and related compounds, paraffin compounds, or aromatic compounds. Most of these compounds have boiling points in excess of 100°C and can be expected to be present, at least initially, as a wood is converted into smoke.

B. Terpenes and Related Compounds

Terpenes and their derivatives represent a diverse group of compounds that have widespread distribution. To date, over 4000 such compounds have been identified, many of which possess potent and characteristic aromas.[21]

Terpenes in wood include both acyclic and cyclic compounds. Terpenes are composed of isoprene (2-methylbutadiene) units with monoterpenes containing two such units; the sesquiterpenes, three; the diterpenes, four; and the triterpenes have six isoprene units. With the addition of each isoprene unit there is the loss of a double bond, and each ring closure also results in the loss of another double bond. The actual positions of the double bonds vary due to isomerization. Also, the corresponding dihydro- and dehydrocompounds can form.

Monoterpenes, and in some species paraffin hydrocarbons, are the major constituents of gum turpentines. The major terpene in turpentine is usually α-pinene, which can exist naturally in the *d* and *l* forms. β-Pinene, also called nopinene, can also exist, but usually only in the *l* form. The bicyclic terpene camphene in both *d* and *l* forms can also be present. Other terpenes found in various wood sources have been summarized in Table 14.

In the case of softwoods, all classes of terpenes from monoterpenes to tri- and tetraterpenes have been found except that relatively few sesterterpenes are present. In contrast, hardwoods contain mainly higher terpenes with some monoterpenes such as camphor being present in certain tropical woods.

All forms of acyclic, monocyclic, and bicyclic monoterpenes have been found in softwoods. Some of the common monoterpenes and their amounts found in various softwood species are summarized in Table 15. In addition, monoterpenes composed of a seven-membered ring have been found in softwoods. These are commonly called tropolone derivatives and are based on the compound 2-hydroxy-2,4,6-cycloheptatrien-1-one.

Softwood oleoresin also contains a relatively high proportion of diterpenes and diterpenoidic acids with the neutral diterpenes being represented by hydrocarbons, oxides, alcohols, and aldehydes. The resin acids, in turn, are mostly tricyclic compounds shown in Table 16. Also, these resin acids can exist as methyl esters.[21]

Most of the softwood triterpenes have a sterane structure and as such can be considered to be steroids. The precursor for most softwood triterpenes is squalene, with the main component of the resulting steroid group in softwoods being β-sitosterol. Most of the triterpenes appear to be esterified with fatty acids.[21]

With hardwoods, sesquiterpenes are relatively rare in woods from the temperate zones but are very common in tropical woods. Relatively few diterpenes are present in hardwoods, however they contain many triterpenes. As in softwoods, β-sitosterol is the major steroid and some of the steroids present in hardwoods are esterified with fatty acids. Also, some tropical hardwoods contain glycosides of triterpenes and steroids which are commonly called saponins, the aglycones of which are called sapogenins.

Some hardwoods contain polyprenes which contain more than six isoprene units. Also, certain hardwoods contain betulaprenols which are alcohols containing six to nine isoprene units and are esterified with saturated fatty acids.

C. Fatty Acids

A fairly wide range of fatty acids and their esters can be found in wood. In most woods the most prevalent are oleic, linoleic, and linolenic acids, although from C_1 to C_{26}, in both, odd and even forms have been found.[10] Most are present as triglycerides although some are in the free form in live wood. However, with seasoning, the triglycerides are hydrolyzed to the free fatty acid form.

Fatty acids are the major components of the parenchyma cell resin in both hardwoods and softwoods. The major saturated fatty acid is palmitic.[21] The relative percentage of free fatty acids is higher in heartwood as compared to sapwood. The normal fat content of most woods is 0.3 to 0.4%.[21]

Waxes, which are esters of fatty acids with higher alcohols, are also present in woods at an average concentration range of 0.08 to 0.09%. The alcohol portion of softwood waxes is usually composed of saturated straight-chain compounds having 16 to 28 carbon atoms. In contrast, both saturated and unsaturated alcohols have been bound in hardwood waxes.

D. Carbohydrates

Aside from the complex carbohydrates already discussed, wood can contain several other carbohydrate forms. For example, simple sugars such as sucrose, glucose, fructose, and arabinose can be present, as well as raffinose, stachyose, and verbascose.[10]

Starch is another carbohydrate form that appears in the sapwood of many hardwoods, and may be present at a level of up to 5%. It appears to act as a reserve food supply.

Another carbohydrate form is pectic substances, which can amount to 0.5% in some woods. They are galacturans associated with arabinans and galactans, and usually occur as the methyl esters of galacturonic acid.

E. Polyhydric Alcohols

Various polyhydric alcohols are also present in wood. The fatty acid glyceride of glycerol is the most common such compound. Others include bergenin, vitexin, keyakinin, keyakinol, pinitol, D-inositol, *myo*-inositol, and sequoyitol.

F. Nitrogen Compounds

There appear to be at least two types of organic nitrogen-containing compounds associated with wood. One class of compounds containing nitrogen is protein and its resulting amino acids. Organic nitrogen-containing compounds can also be represented in wood by alkaloids. Not all woods contain alkaloids. Typical wood alkaloids include berberine, liriodenine, anibine, dictamnine, and maculine.

G. Phenolic Compounds

Numerous phenolic-based compounds are also present in certain wood species. These represent a wide range of compound types including acids, aldehydes, alcohols, lignans, stilbenes, tannins, and various types of flavonoids. Some of the major compounds are summarized in Table 17.

At this point some of the various phenolic-based compound classes present in wood will be discussed. The first general category is the simple phenols characterized by compounds such as vanillin, syringin, coniferin, and *p*-hydroxybenzaldehyde. Many of these compounds appear to be residues and/or by-products of lignin biosynthesis.[21]

Lignans represent another phenolic group of compounds. They are composed of two phenylpropane units linked in a different manner. Numerous types of linkages exist including β–β, α–γ, tetrahydrofuran ring, and dihydrofuran ring. Interestingly, some of these compounds have dimeric structures which are also part of the intact lignin molecule. Lignans appear to be primarily associated with heartwood with the amount present in sapwood being quite small.

Another group is the stilbenes which are primarily associated with the heartwood of pines. A conglomerate of compounds commonly called flavonoids and composed of flavones, flavanes, flavanones, isoflavones, chalcones, and aurones have also been identified in various wood species.[21]

Tannins, which can be further subdivided into hydrolyzable and nonhydrolyzable or condensed (phlobaphenes) tannins are also present in wood with the hydrolyzable ones occurring more frequently. Hydrolyzable tannins are esters of gallic acid and its dimers whereas condensed tannins are primarily derived from three to eight units composed of catechins and leucoanthocyanidins.

H. Inorganic Constituents

The mineral content of wood is dependent upon the soil conditions and thus, the same species can have vastly different mineral composition. Most minerals exist as calcium, potassium, and magnesium salts. Numerous other trace metals are also present.

Woods from temperate zones are usually relatively low in ash (0.1 to 1.0%) whereas tropical woods can have up to 5% ash. Typical inorganic composition of various softwoods and hardwoods are shown in Table 18.

VI. BARK COMPOSITION

In certain instances bark may represent a readily available by-product that can serve as an inexpensive wood smoke source. It usually represents 10 to 15% of total tree weight and has a composition that is much more complex than wood.[16]

Bark can be divided into living inner bark, called phloem, and dead outer bark, called rhytidome. Inner bark is composed of sieve elements, parenchyma cells, and schlerenchymatous cells while outer bark is primarily composed of periderm or cork layers.

Jensen et al.,[11] as seen in Table 19, have clearly demonstrated significant differences in the composition of bark vs. wood, as well as differences between inner and outer barks. When outer bark and wood are compared, it can be seen that bark is higher in protein, suberin, and fat than wood but much lower in lignin.

Suberin is the major constituent of cork cell walls and can be found at levels up to 39% in various species of birch bark while most common pine and oak species have 2 to 8% suberin.[11] The suberin content in the outer bark of the cork tree is 35 to 40% and is the commercial source of cork. Compositionally, suberin is mainly an ester of aliphatic hydroxy acids and various phenolic acids. Actually, many forms of suberin exist since various bonds can be formed. Besides forming ester bonds with monobasic acids, they can contain α,β–dibasic acids that are esterified with diols. There are also various double bonds and hydroxyl groups that interact to form ester and ether cross-linkages. As a result, their chain lengths can vary but most suberins contain 16 to 18 carbon atoms. The composition of suberin is influenced by bark species, as seen in Table 20. In addition, cutin, which has a structure similar to suberin but is heavily branched, is also present in the cork fraction.

Bark can also be chemically divided into soluble constituents or extractives, and insoluble constituents. In turn, bark extractives can be subdivided into lipohilic and hydrophilic components. Compared to wood, bark extractives are relatively high, representing 20 to 40% of bark dry weight. The lipophilic fraction is composed of fats, waxes, terpenes, terpenoids, and higher aliphatic alcohols while the hydrophilic fraction contains numerous types of phenolic constituents. In addition, small amounts of soluble carbohydrates, including starch, pectins, raffinose, and stachyose, as well as proteins and vitamins, are present in the hydrophilic portion.

Bark also contains 2 to 5% inorganic solids or ash,[16] but some species can contain upwards of 10%.[17] Calcium and potassium are the major minerals present in the form of various salts including oxalates, phosphates, and silicates. Minor trace elements include boron, copper, and manganese. However, as seen in Table 21, the mineral composition of bark can vary significantly with species.

The composition of bark has been investigated and a wide variety of compounds identified, some of which are apparently unique to bark. Some of the compounds identified are summarized in Table 22.

The major sugar present from the hydrolysis of bark is glucose, which is derived from cellulose, and as seen in Table 23, can range from 16 to 41%, with the outer bark having less glucose than the inner bark. Actual carbohydrate composition of bark is primarily cellulose, as seen in Table 24, but numerous polyoses, or hemicelluloses, as seen in Table 25, have been identified.

The polyphenols in bark can be classified relative to their molecular weight and solubility. Procyanidins, which are essentially di- and trimeric flavanols, are the smallest molecular weight. The next type are the condensed tannins or phlobaphenes, that range from 1000 to 2000 in molecular weight. As seen in Table 26, bark-condensed tannins range from 2 to 54%. The third major type of polyphenol is polyphenolic acids which range in molecular weight from 1500 to 6700. Normally they contain 1 to 4% aliphatic hydroxy groups with less than 2% methoxy groups.

Because of the presence of a significant amount of acidic components in bark, the pH of bark is usually more acidic than the corresponding wood with the outer bark being more acidic than inner. Typical pH values are summarized in Table 27.

Various pigments, commonly called flavonoids, but including the specific compounds flavone, isoflavone, flavonone derivatives, anthocyanins, and leucoanthocyanins, are present in bark. Derivatives of flavone, isoflavone, and flavonone contribute yellow color and are derived from 2- or 3-phenylbenzo-γ-pyrone.[11]

All bark usually contains protein since bark serves as its storage site. In birch, inner bark contains approximately 5% protein while outer bark is lower at 3.8%[11] however, protein content can vary throughout the year. Protein values as high as 21.6% have been reported,[11] but since this value was calculated from percent nitrogen, its accuracy is questionable since bark can contain many nitrogen-containing compounds. In addition to intact protein, bark can also contain free amino acids.[11]

The bark of angiosperms contains alkaloids whereas dicotyledon bark seldom contains alkaloids.[11] Usually, alkaloids are not present in free form but form salts with compounds such as citric, tannic, oxalic, and phosphoric acids. Typical alkaloids found in bark are summarized in Table 28. Bark also contains vitamins whose concentrations are seasonal and species dependent. Data for springtime barks are summarized in Table 29.

VII. HARDWOODS VS. SOFTWOODS

The above discussions have addressed wood composition in general and obviously there are significant differences dependent upon species. Perhaps it would be helpful to become somewhat more specific by looking at compositional differences between softwoods and hardwoods, since it is well established that smoke derived from these two broad categories are compositionally and organoleptically different.

Structurally, hardwoods can be considerd to be more complex than softwoods. Hardwoods in general contain more cell types and show a greater variation in size, shape, and arrangement as compared to softwoods.

A total of seven different types of softwood cells exist, although in most softwoods only two are predominant, namely longitudinal tracheids and ray parenchyma cells. Longitudinal tracheids are by far the most abundant and participate in conduction and support while the ray parenchyma cells are used to store food.

In the case of hardwoods, a total of ten different types of cells have been identified, however, four types of cells make up most of the structure of hardwood. These include vessel elements, fibers, longitudinal parenchyma, and ray parenchyma. In hardwoods, the vessel elements are used for conduction while the fibers function for support.

In general, it can be stated that the hemicellulose fraction of hardwoods is higher in pentosans than softwoods, while in softwoods, hexosans predominate.[12] As will be seen in a later chapter, this results in different end products during smoking.

Major compositional differences in lignin also exist between hardwoods and softwoods in that there are approximately three more syringylpropane units than guaiacylpropane units in hardwood as in softwood, and as a result, as shown in Table 30, structural differences result.

The general quantitative differences among cellulose, lignin, and pentosan hemicellulose between hardwoods and softwoods are perhaps best typified by the data presented in Table 31. As can be seen, cellulose amounts are fairly constant independent of species, while lignin is higher in softwoods, and pentosan hemicellulose is higher in hardwoods.

Other carbohydrate differences are that the mannan content of most softwoods is in the 10 to 15% range, while for hardwoods, the mannan amount is rarely over 3% (Table 32). However, the xylan content of hardwoods is 12 to 20% while in softwoods, it seldom exceeds 10%.[2]

Another significant compositional difference is the fact that softwoods contain more resin acids and other extractives that are soluble in organic solvents than hardwoods. Also, most hardwoods have very low levels of volatiles, while most softwoods can have fairly high amounts.

Within hardwood species, there are compositional differences due to tropical or temperate site of growth. Hardwoods from the tropics are higher in extractives and ash, while they are lower in lignin and acetyl groups than their counterparts grown in temperate zones.

VIII. HEARTWOOD VS. SAPWOOD

The ratio of heartwood to sapwood can vary dramatically with different woods and even within species, with some species not producing a true sapwood. Also, the proportion of sapwood generally increases with tree height. Rate of tree growth also is an important factor with faster growing trees having less heartwood. Environmental factors such as a favorable climate and soil conditions result in the formation of a smaller proportion of heartwood. From a compositional standpoint, heartwood is lower in moisture than sapwood.

With softwoods the heartwood, or center portion, has less lignin and cellulose but more extractives than sapwood, while in hardwoods no consistant differences are apparent.[2,14] In both hardwoods and softwoods the acetyl content[2] and the acetic acid content[14] are higher in sapwood than in hardwood. No conclusive differences have been found for ash.[14]

Table 1
TYPICAL AMERICAN HARDWOOD AND SOFTWOOD SPECIES

Hardwoods	Softwoods
Acer macrophyllum — Bigleaf maple	*Abies amabilis* — Pacific silver fir
Acer negundo — Boxelder	*Abies balsamea* — Balsam fir
Acer rubrum — Red maple	*Abies concolor* — White fir
Acer saccharinum — Silver maple	*Abies lasiocarpa* — Subalpine fir
Alnus rubra — Red alder	*Abies procera* — Noble fir
Arbutus menziesii — Pacific madrone	*Chamaecyparis thyoides* — Atlantic white cedar
Betula alleghaniensis — Yellow birch	*Juniperus deppeana* — Alligator juniper
Betula nigra — River birch	*Larix larcina* — Tamarack
Betula papyrifera — Paper birch	*Larix occidentalis* — Western larch
Carya cordiformus — Bitternut hickory	*Libocedrus decurrens* — Incense cedar
Carya glaubra — Pignut hickory	*Picea engelmanni* — Englemann spruce
Carya ovata — Shagbark hickory	*Picea glauca* — White spruce
Carya pallida — Sand hickory	*Picea mariana* — Black spruce
Carya tomentosa — Mockermut hickory	*Picea sitchensis* — Sitka spruce
Celtis laevigata — Sugarberry	*Pinus attenuata* — Knobcone pine
Eucalyptus gigantea — Eucalyptus	*Pinus banksiana* — Jack pine
Fagus grandifolia — American beech	*Pinus clausa* — Sand pine
Fraxinus americana — White ash	*Pinus contorta* — Lodgepole pine
Fraxinus pennsylvanica — Green ash	*Pinus echinata* — Shortleaf pine
Gleditsia triacanthos — Honey locust	*Pinus elliottii* — Slash pine
Laguncularia racemosa — White mangrove	*Pinus monticola* — Western white pine
Liquidambar styraciflua — Sweetgum	*Pinus palustris* — Longleaf pine
Liriodendron tulipifera — Yellow poplar	*Pinus ponderosa* — Ponderosa pine
Lithocarpus densiflorus — Tan oak	*Pinus resinosa* — Red pine
Milalenca quinquenervia — Cajeput	*Pinus sabiniana* — Digger pine
Nyssa aquatica — Water tupelo	*Pinus strobus* — Eastern white pine
Nyssa sylvatica — Black tupelo	*Pinus taeda* — Loblolly pine
Populus alba — White poplar	*Pseudotsuga menziesii* — Douglas fir
Populus deltoides — Eastern cottonwood	*Sequoia sempervirens* — Redwood
Populus tremoides — Quaking aspen	*Taxodium distichum* — Bald cypress
Populus trichocarpa — Black cottonwood	*Thuja occidentalis* — Northern white cedar
Prunnus serotina — Black cherry	*Thuja plicata* — Western red cedar
Quercus alba — White oak	*Tsuga canadensis* — Eastern hemlock
Quercus coecinea — Scarlet oak	*Tsuga heterophylla* — Western hemlock
Quercus douglasii — Blue oak	*Tsuga mertensiana* — Mountain hemlock
Quercus falcata — Southern red oak	
Quercus kelloggii — California black oak	
Quercus lobata — Valley oak	
Quercus lyrata — Overcup oak	
Quercus marylandica — Blackjack oak	
Quercus prinus — Chestnut oak	
Quercus rubra — Northern red oak	
Quercus stellata — Post oak	
Quercus velutina — Black oak	
Salix nigra — Black willow	
Tila heterophylla — Basswood	
Ulmus americana — American elm	
Ulmus crassifolia — Cedar elm	

Table 2
STRUCTURAL INTERRELATIONSHIPS OF THE MAJOR COMPONENTS OF WOOD

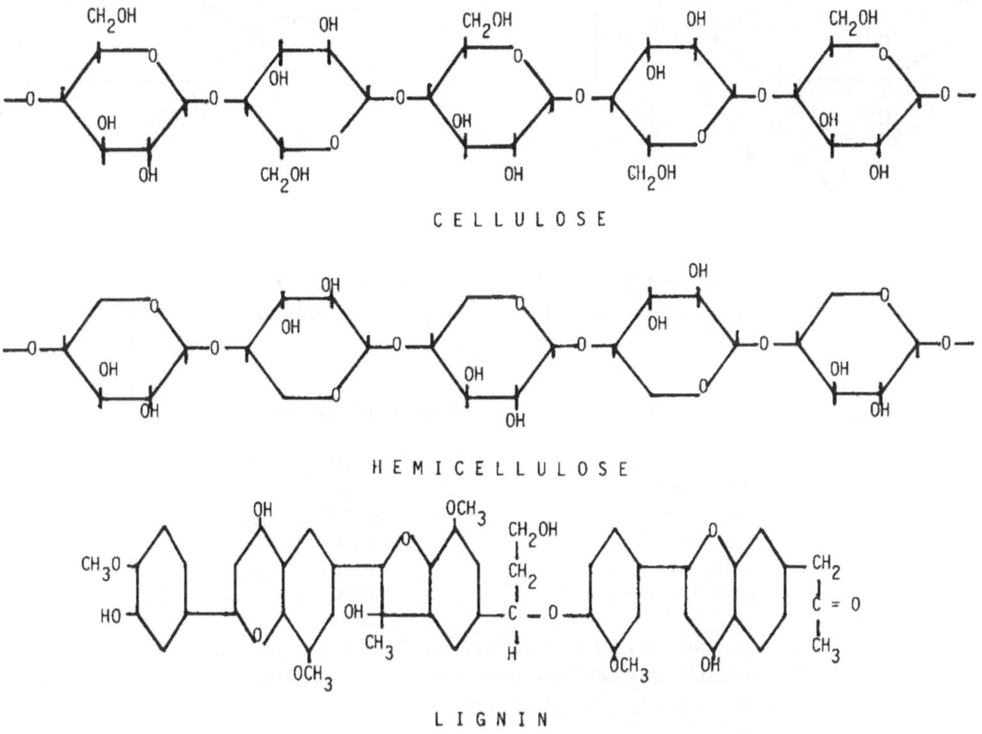

Table 3
INTERRELATIONSHIPS OF WOOD COMPONENTS

- Wood
  - Major Cell Wall Components
    - Lignin
    - Polysaccharides
      - Cellulose
        - Hydrolysis
          - D-glucose
      - Hemicelluloses
        - Hydrolysis
          - Hexoses
          - Pentoses
          - Uronic Acids
          - Methoxyuronic Acids
          - Acetyl Groups
  - Minor Components
    - Soluble or Volatile
      - Extractives
      - Fats
      - Resins
    - Insoluble
      - Inorganics
      - Protein
      - Pectins

Table 4
STRUCTURAL RELATIONSHIPS OF THE END GROUPS OF CELLULOSE

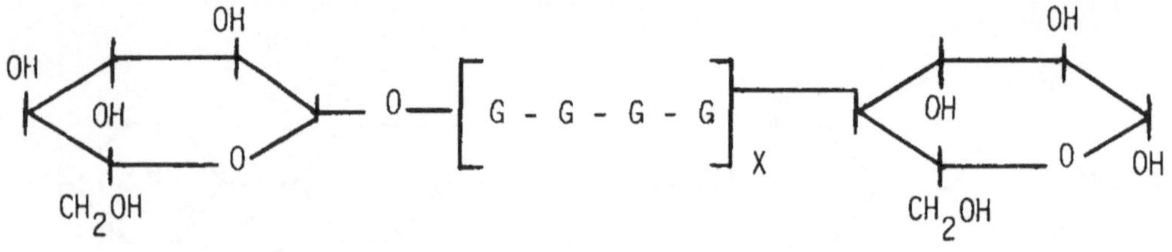

(Nonreducing) (Reducing)

Table 5
DEGREES OF POLYMERIZATION OF CELLULOSE AMONG VARIOUS WOOD SOURCES

Product	Degree of polymerization
Trembling aspen wood	10,300
White birch wood	9,400
Englemann spruce wood	8,000
Jack pine wood	7,900
White birch bark	7,500
Englemann spruce bark	7,100

Adapted from Fengel, D. and Wegener, G., in *Wood: Chemistry, Ultrastructure, Reactions*, Walter de Gruyter, Berlin, 1983, chap. 4.

Table 6
STRUCTURE RELATIONSHIPS OF HEMICELLULOSE COMPONENTS

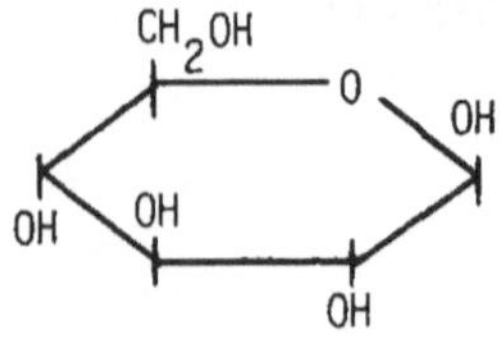

Glucose

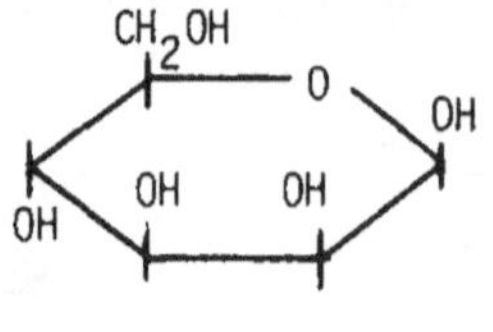

Mannose

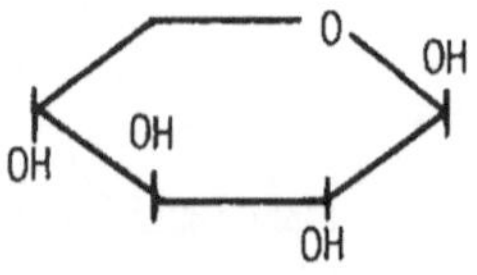

Xylose

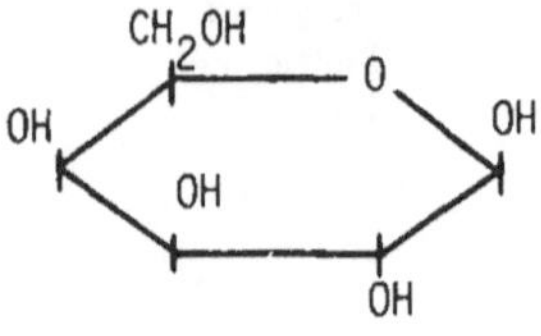

Galactose

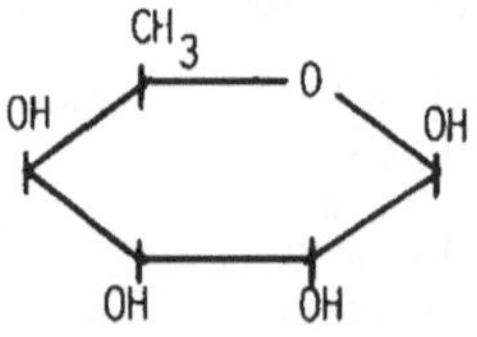

Rhamnose

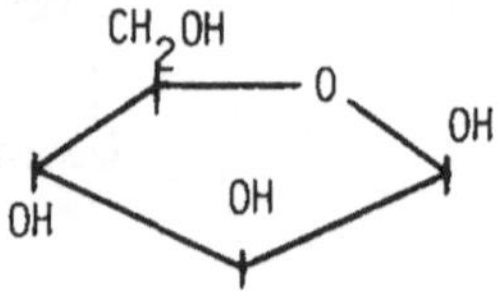

Arabinose

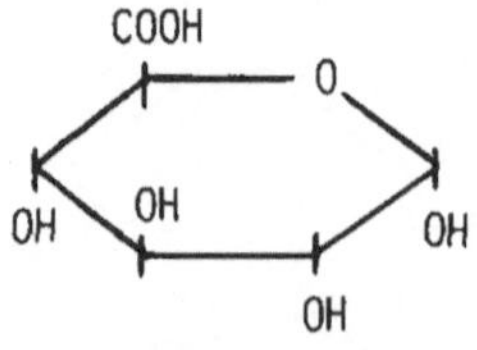

Glucuronic Acid

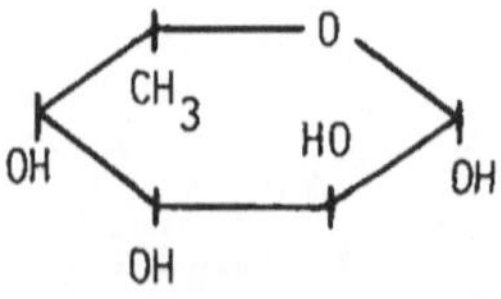

Fucose

Table 7
PROPERTIES OF MAJOR WOOD HEMICELLULOSES

Hemicellulose	% in wood	Units	Molar ratios	Linkage	Units/chain
Galactoglucomannan	5—8	β-D-Mannose	3	1—4	100
		β-D-Glucose	1	1—4	
		α-D-Galactose	1	1—6	
(Galacto)Glucomannan	10—15	Acetyl			
		β-D-Mannose	4	1—4	100
		β-D-Glucose	1	1—4	
		α-D-Galactose	0.1	1—6	
		Acetyl	1		
Arabinoglucuronoxylan	7—10	β-D-Xylose	10	1—4	100
		4-*O*-Methylglucuronic acid	2	1—2	
Arabinogalactan	5—35	β-D-Galactose	6	1—3, 1—6	200
		α-L-Arafinose	0.6	1—6	
		β-L-Arafinose	0.3	1—3	
		β-D-Glucuronic acid	trace	1—6	
Glucuronoxylan	15—30	β-D-Xylose	10	1—4	200
		4-*O*-Methylglucuronic acid	1	1–	
		Acetyl			
Glucomannan	2—5	β-D-Mannose	1—2	1—4	200
		β-D-Glucose		1—4	

Adapted from Pettersen, R. C., in *The Chemistry of Solid Wood,* Rowell, R., Ed., American Chemical Society, Washington, D.C., 1984, chap. 2.

Table 8
STRUCTURE OF *O-ACETYL-4-O*-METHYLGLUCURONOXYLAN

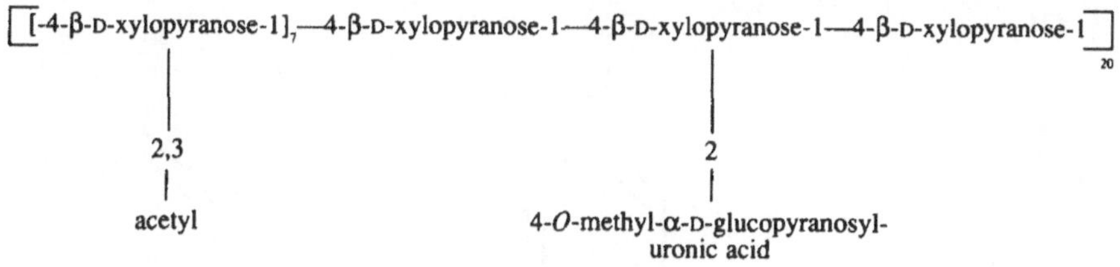

Table 9
COMMON XYLANS IN VARIOUS WOODS

Wood	Xylan	Ratio	Branched	Units/chain
Yellow birch	Xylose/2-α-(4-*O*-Methyl Gu)	10:1	No	192
Finnish birch	Xylose/2-α-(4-*O*-Methyl Gu)	22:1	No	22
American beech	Xylose/2-α-(4-*O*-Methyl Gu)	8:1	One at C_2	45
European beech	Xylose/2-α-(4-*O*-Methyl Gu)	10:1	No	70
Sugar maple	Xylose/2-α-(4-*O*-Methyl Gu)	10:1	No	205
White elm	Xylose/2-α-(4-*O*-Methyl Gu)	7:1	No	211
Trembling aspen	Xylose/2-α-(4-*O*-Methyl Gu)	9:1	No	200
European larch	Xylose/2-α-(4-*O*-Methyl Gu) 3Ar	6:1:1	No	100
Spruce	Xylose/2-α-(4-*O*-Methyl Gu) 3Ar	7:1:1	No	95
Scots pine	Xylose/2-α-(4-*O*-Methyl Gu) 3Ar	7.5:1.5:1	No	133

Adapted from Schuerch, C., in *The Chemistry of Wood,* Browning, B. L., Ed., R. E. Krieger, Huntington, N.Y., 1975, chap. 5.

Table 10
LIGNIN CONTENT OF VARIOUS WOODS

Wood	Lignin (%)
Aspen	17.6
Birch	21.0
English oak	22.4
Maple	23.2
Beech	23.5
Ash	24.8
Norway spruce	26.1
White spruce	26.5
Sitka spruce	27.8
Pitch pine	29.2
Balsam fir	30.1
Eastern hemlock	31.5
African pencil cedar	37.7

Adapted from Sarkanen, K. V., in *The Chemistry of Wood*, Browning, B. L., Ed., R. E. Krieger, Huntington, N.Y., 1975, chap. 6.

Table 11
ALCOHOLS INVOLVED IN THE FORMATION OF LIGNIN

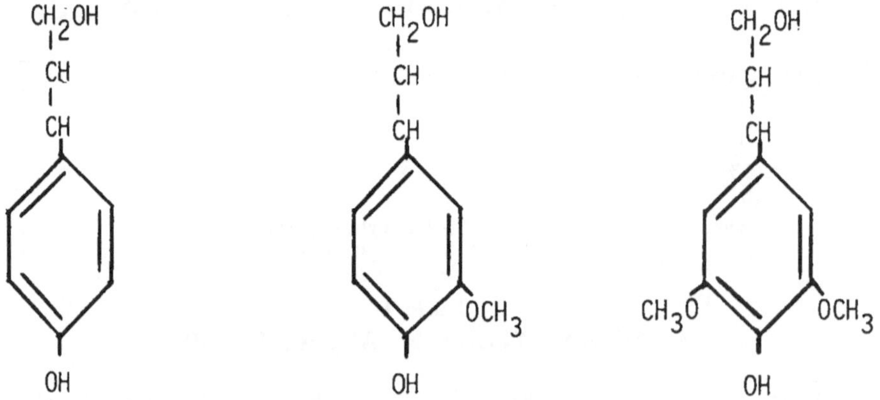

p-Coumaryl Alcohol — Coniferyl Alcohol — Sinapyl Alcohol

Adapted from Freudenberg, K., *Science*, 148, 595, 1965.

Table 12
COMMON LIGNIN LINKAGES AND THEIR RELATIVE AMOUNTS IN SPRUCE AND BIRCH

	% in	
Linkage	Spruce	Birch
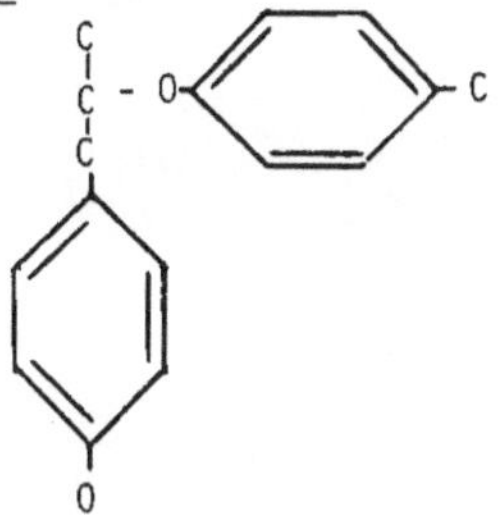	48	60
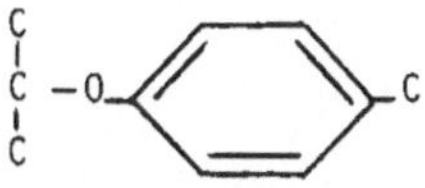	2	2
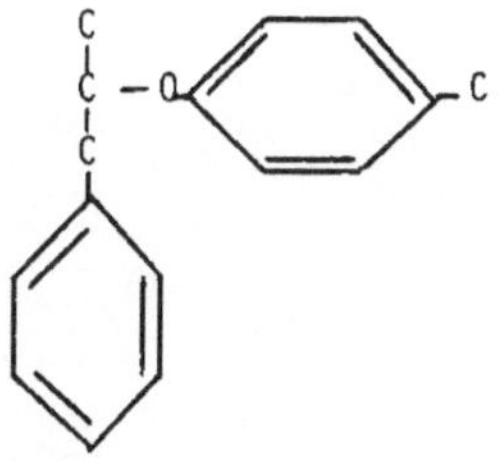	7	7
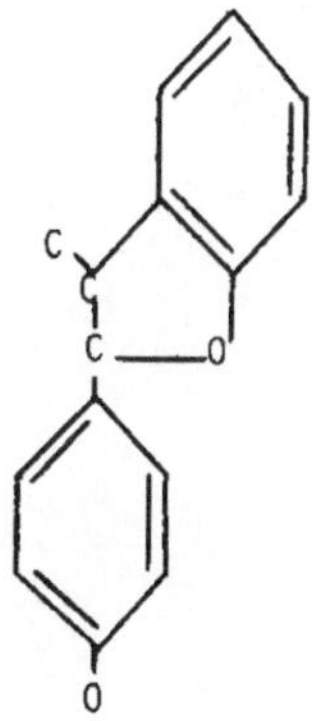	9—12	6
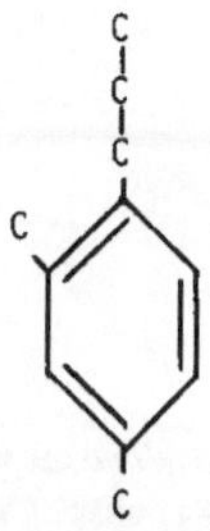	2.5—3	1.5—3.0

Table 12 (continued)
COMMON LIGNIN LINKAGES AND THEIR RELATIVE AMOUNTS IN SPRUCE AND BIRCH

	% in	
	Spruce	Birch
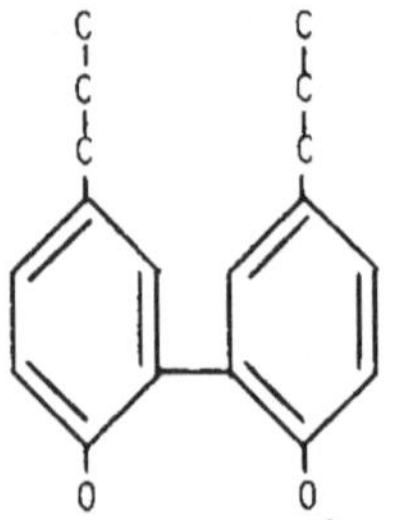	9.5—11	4.5
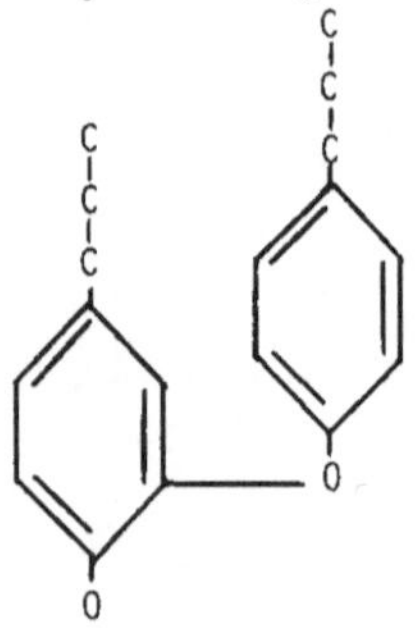	3.5—4	6.5
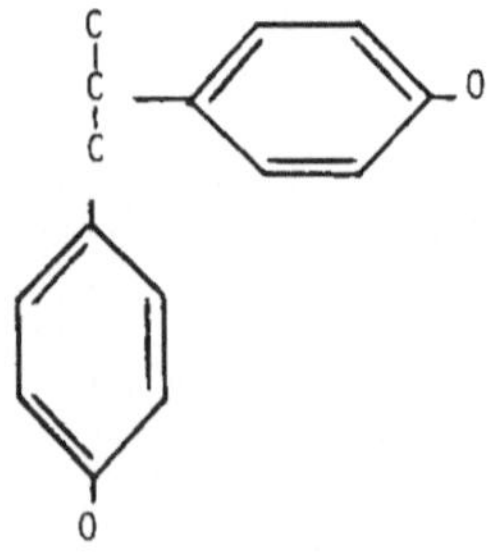	7	7
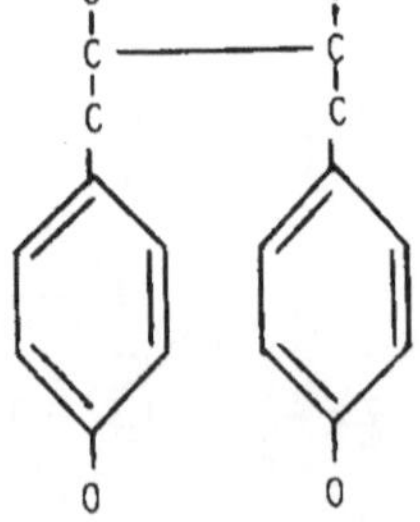	2	3

Adapted from Goldstein, I. S., in *The Chemistry of Solid Wood,* Rowell, R., Ed., American Chemical Society, Washington, D.C., 1984, chap. 15.

Table 13
EXTRACTIVES COMPOSITION OF VARIOUS WOODS (%)

	Picea abies	*Pinus sylvestris*	*Betula verrucosa*	*Populus tremula*
Acetone extract	2.22	3.10	3.46	4.53
Petroleum ether solubles	1.04	2.29	2.03	2.27
Free fatty acids	7.52	7.55	—	4.67
Resin acids	27.37	29.03	0.30	0.75
Neutral compounds	62.13	60.59	99.70	94.58
Hydrocarbons	1.84	3.21	1.99	5.01
Waxes	8.41	3.64	9.67	13.53
Triglycerides	18.67	39.17	39.88	45.68
Higher alcohols	9.55	6.06	7.68	10.59
Diglycerides	5.26	1.39	8.48	2.55
Monoglycerides	5.26	0.79	11.07	5.20
Oxidized compounds	13.05	1.89	14.46	6.72

Adapted from Fengel, D. and Wegener, G., in *Wood: Chemistry, Ultrastructure, Reactions*, Walter de Gruyter, Berlin, 1983, chap. 7.

Table 14
MISCELLANEOUS TERPENES IDENTIFIED IN VARIOUS WOODS

Δ^3-Carene	Cadinene	Dipterocarpol
Δ^4-Carene	Longifolene	Cycloartenol
Sylvestrene	Cedrene	Cycloeucalenol
Limonene	Cedrenol	α- and β-Amyrin
β-Phellandrene	Cedrol	Ferreol
α-Thujene	α- and β-Santalene	Castangogenin
Santene	α- and β-Santaiol	Oleanolic acid
Terpinolene	α- and β-Gurjunene	Arjenolic acid
α-Terpinene	Lanceol	Termiolic acid
Ocimene	Cadinol	Morlolic acid
Myrcene	Guaiol	Butulinic acid
Linalool	α- and β-Atlantone	Emmolic acid
α-Terpineol	Cadinene	Dihydropalustric acid
Dihyro-α-terpineol	Cedrene	
Fenchyl alcohol	Azulene	β-Sitosterol
Borneol	Cuparene	Squalene
Camphor	Camphorene	
Tropolene	Phytol	
α-Thujaplicin	Ferruginol	
β-Thujaplicin	Totarol	
γ-Thujaplicin	Hinokiol	
β-Thujaplicinol	Manoyloxide	
β-Dolabrin	Ketomanoyloxide	
Nootkatin	Abietic acid	
Chamaecin	Heyderiol	
Pyqmaein	Esdragol	
p-Cymene	Pimaric acid	
Carvacrol	Retene	
Thymol	Pimanthrene	
p-Methoxythymol	Levopimaric acid	
p-Methoxycarvacrol	Neoabietic acid	
Hydrothymoquinone	Dehydroabietic acid	
Thymoquinone	Isopimaric acid	
Libocedrol	Dihydroabietic acid	

Adapted from Buchanan, M. A., in *The Chemistry of Woods*, Browning, B. L., Ed., R. E. Krieger, Huntington, N.Y., 1975, chap. 7.

Table 15
MONOTERPENE COMPOSITION OF VARIOUS SOFTWOODS

	Monoterpene (%)									
Wood	**1**	**2**	**3**	**4**	**5**	**6**	**7**	**8**	**9**	**10**
Abies alba	39.0	3.0	4.5	—	53.5	—	—	—	—	—
A. amabilis	15.5	16.5	38.0	0.5	1.0	—	—	—	1.0	26.5
A. grandis	13.0	18.5	0.5	34.5	5.0	—	—	—	0.5	26.0
Picea abies	58.0	24.0	2.1	3.0	4.5	0.5	—	0.4	—	—
Pinus aristata	8.8	11.2	64.4	0.1	7.0	—	—	3.4	3.6	2.3
P. balflouriana	81.0	1.9	—	0.9	14.1	0.3	—	1.0	0.6	0.2
P. elliottii	62.6	20.6	—	1.4	1.7	—	—	0.4	1.7	8.1
P. heldreichii	11.0	0.5	—	0.2	82.8	—	—	—	0.6	—
P. monophylla										
Sapwood	46.0	2.0	3.0	1.0	7.0	—	5.0	4.0	22.0	8.0
Heartwood	67.0	2.0	—	1.0	3.0	—	1.0	1.0	20.0	3.0
P. occidentalis	63.8	22.2	7.7	0.2	—	—	—	—	1.1	—
P. ponderosa	10.2	16.5	36.3	1.4	12.1	1.0	1.4	1.9	—	1.4
P. strobus	67.0	18.0	—	2.9	0.9	—	—	0.5	0.9	0.5
P. taeda	64.0	28.0	—	1.3	1.5	—	—	—	2.0	0.5
P. tropicalis	93.2	3.1	—	1.2	0.8	—	—	0.1	0.8	0.6

Note: 1. α-Piene, 2. β-Piene, 3. Δ^3-Carene, 4. Camphene, 5. Limonene, 6. *p*-Cymene, 7. Terpinene, 8. Terpinolene, 9. Myrcene, 10. β-Phellandrene.

Adapted from Fengel, D. and Wegener, G., in *Wood: Chemistry, Ultrastructure, Reactions,* Walter de Gruyter, Berlin, 1983, chap. 7.

Table 16
DITERPENE RESIN ACID COMPOSITION OF VARIOUS SOFTWOODS

	Diterpene resin acid (%)									
Wood	**1**	**2**	**3**	**4**	**5**	**6**	**7**	**8**	**9**	**10**
Picea abies										
Sapwood	3.3	5.6	17.5	28.6	11.3	—	9.3	13.2	11.0	—
Heartwood	4.4	6.4	16.4	10.5	11.8	—	8.2	35.7	6.5	—
Pinus banksiana										
Sapwood	6.0	2.0	13.0	1.0	6.0	—	15.0	51.0	3.0	—
Heartwood	7.0	2.0	11.0	—	8.0	—	29.0	35.0	4.0	—
P. elliottii	7.4	2.0	19.4	9.6	23.7	—	12.0	4.8	18.4	
P. palustris	6.0	1.8	15.2	27.9	5.0	—	15.1	15.3	13.7	—
P. pinaster	10.0	1.0	50.0	—	4.1	—	15.0	4.2	16.0	—
P. ponderosa										
Sapwood	9.0	2.0	10.0	5.0	15.0	—	17.0	31.0	8.0	—
Heartwood	9.0	2.0	20.0	2.0	10.0	—	28.0	15.0	13.0	—
P. strobus										
Sapwood	1.0	2.0	8.0	—	34.0	14.0	29.0	5.0	5.0	—
Heartwood	1.0	2.0	8.0	—	36.0	11.0	32.0	5.0	3.0	—
P. sylvestris	8.1	1.6	15.1	30.0	3.5	—	15.8	14.4	11.1	0.4

Note: 1. Pimaric, 2. Sandaracopimaric, 3. Palustric, 4. Levopimaric, 5. Isopimaric, 6. Anticopalic, 7. Abietic, 8. Dehydroabietic, 9. Neoabietic, 10. Secodehydroabietic.

Adapted from Fengel, D. and Wegener, G., in *Wood: Chemistry, Ultrastructure, Reactions,* Walter de Gruyter, Berlin, 1983, chap. 7.

Table 17
PHENOLIC-BASED COMPOUNDS IDENTIFIED IN WOOD

Phthalic acid	Chrysin	Tannins
Piperonylic acid	Morin	Maclurin
Gallic acid	Fisetin	Cotoin
Ellagic acid	Robinetin	Anibine
p-Hydroxybenzoic acid	Quercetin	Euxanthone
Vanillic acid	Kaempferol	Brazilin
Syringic acid	Pinocembrin	Haematin
Ferulic acid	Naringenin	Lapachol
Vanillin	Taxifolin	Desoxysantalin
Syringaldehyde	Aromadendrin	Chrysarobin
Coniferin	Dihydromorin	Tectoquinone
Syringin	Dihydrorobinetin	
Matairlesinol	Pinistrobin	
Lariciresinol	Pinobanksin	
Pinoresinol	Strobopinin	
Conidendrin	Cryptostrobin	
Secoisolariciresinol	Strobobanksin	
Livoviol	Tectochysin	
Oxomatairesinol	Afzelin	
Hydroxymatairesinol	Ayanin	
Allo-hydroxymatairesinol	Santal	
Pinosylvin	Prunetin	
4-Hydroxystilbene	Muningin	
4-Methoxystilbene	Podospicatin	
2,4,3′,5′-Tetrahydroxystilbene	Pterocarpin	
Okanin	Homopterocarpin	
Isookanin	Catechin	
Neoplathymenin	Leucofisentinidin	

Adapted from Buchanan, M. A., in *The Chemistry of Wood*, Browning, B. L., Ed., R. E. Krieger, Huntington, N.Y., 1975, chap. 7.

Table 18
INORGANIC COMPOSITION (PPM) OF VARIOUS SOFTWOODS AND HARDWOODS

	Ca	K	Mg	Mn	Na	Cl	P	Al	Fe	Zn
Softwood										
Abies balsamea	830	770	270	127	—	—	—	—	13	11
Picea rubens	820	200	70	144	—	—	50	—	14	8
Pinus trobus	210	290	70	28	—	—	—	—	10	11
Pseudotsuga menziesii	295	—	41	25	44	67	—	13	—	—
Thuja plicata	1013	229	76	1	10	12	—	2	—	—
Hardwood										
Acer rubrum	820	690	120	72	—	—	30	2	11	29
A. saccharum	1064	990	140	36	29	82	—	1	—	—
Betula papyrifera	740	270	180	34	—	—	150	23	10	28
Quercus alba	674	780	11	2	3	8	—	6	—	—
Tilia americana	1125	543	117	11	74	93	—	15	—	—

Adapted from Fengel, D. and Wegener, G., in *Wood: Chemistry, Ultrastructure, Reactions*, Walter de Gruyter, Berlin, 1983, chap. 7.

Table 19
COMPOSITIONAL DIFFERENCES AMONG OUTER, INNER BARK, AND BIRCH WOOD

Fraction	Outer bark (%)	Inner bark (%)	Wood (%)
Ash	0.4	1.8	0.3
Protein	3.8	5.0	1.9
Lignin	1.3	20.3	19.5
Pentosans	1.1	20.2	25.2
Hexosans	3.4	18.5	43.4
Reducing sugars	—	1.3	—
Suberin	38.7	1.2	—
Coloring matter, tannins	5.6	13.7	1.8
Fats, waxes, resin acids	38.1	1.7	1.3
Acetyl groups	0.24	2.8	4.6
Formyl groups	0.04	0.1	0.2
Unidentified	7.3	13.4	1.6

Adapted from Jensen, W., Fremer, K. E., Sierila, P., and Wartiovaara, V., *The Chemistry of Wood,* Browning, B. L., Ed., R. E. Krieger, Huntington, N.Y., 1975, chap. 12.

Table 20
SUBERIN COMPOSITION OF VARIOUS BARKS

Composition	% in	
	Birch	Oak
Neutral		
Alkan-l-ols	—	2.7
Unidentified	10.3	3.4
Acidic		
Monobasic	—	1.9
α, ω,-Dibasic	8.5	7.6
ω-Hydroxymonobasic	21.3	47.4
Dihydroxyoctadecanoic	1.3	—
Dihydroxyhexadecanoic	3.6	—
9,10,Dihydroxyoctadecane-1,18-dioic	1.3	15.4
9,10,18-Trihydroxyoctadecanoic	42.7	7.7
9,10-Epoxy-18-hydroxyoctadecanoic	1.8	—
Unidentified	9.2	13.9

Adapted from Fengel, D. and Wegener, G., *Wood: Chemistry, Ultrastructure, Reactions,* Walter de Gruyter, Berlin, 1983, chap. 9.

Table 21
MINERAL CONTENT OF THE BARK FROM VARIOUS WOODS

Species	% Ash	% Ca	% K	% Mg	% Na	% P	% Mn	% Zn
Willow								
Outer bark	11.5	88.80	6.41	2.24	0.82	0.53	0.79	0.39
Inner bark	13.1	81.85	14.01	1.31	0.91	1.08	0.53	0.30
Red oak								
Outer bark	8.9	92.06	3.45	2.58	0.31	0.54	0.97	0.09
Inner bark	11.1	95.00	2.58	1.24	0.26	0.51	0.34	0.07
Ash								
Outer bark	12.3	90.38	5.37	3.08	0.22	0.45	—	0.50
Inner bark	12.1	82.44	12.90	2.81	0.23	1.21	—	0.40

Adapted from Fengel, D. and Wegener, G., *Wood: Chemistry, Ultrastructure, Reactions*, Walter de Gruyter, Berlin, 1983, chap. 9.

Table 22
COMPOUNDS IDENTIFIED IN WOOD BARK

Lignoceryl alcohol	Galbulin	Chinovic acid
Behenyl alcohol	Galacatin	Phyllanthol
D-Mannite	Galbacin	Alnusenone
Dulcite	Galgravin	Melaleucic acid
d-Quercitol	Eudesmin	Taxifolin
Quebrachitol	Asarinn	Dihydromyricetin
Inositol	Sympiolosin	Aromadendrin
Monotropitoside	Liriodendrin	Pinobanksin
Piperonylic acid	Plumbagin	Quercetin
p-Coumaric acid	Alkannan	Myricetin
Caffeic acid	Alkannin	Delphinidin
Fagaramide	Anthragallol	Delargonindin
Protocatechuic acid	Rubiadin	Cyanidn
Ellagic acid	Suranjidiol	Leucodelphinidin
Cinnamaldehyde	Morindone	Berberine
Salicylic aldehyde	Morindin	Canadine
4-Methoxysalicylic aldehyde	Emodin	Hydrastine
Myristic aldehyde	Faxetin	Berbamine
Gluovanillin	Fraxin	Quinine
Picein	Esculetin	Cinchonine
Cotion	Esculin	Quinidine
Hydrocotioin	Umbelliferone	Aricine
Maclurin	Scopoletin	Conessine
Protocotoin	Xanthoxyletin	Kurchine
Phloretin	Maltol	Pelletierine
d-Camphor	Rotenone	Pseudopelletierne
Salicin	Betulinol	Strychnine
Populin	Betulinic acid	Brucine
Tremuloidin	Freidelin	Yohimbine
Salireposide	Cerin	Aspidospermine
Betuloside	Taraxerol	Rauwolfine
Coniferin	Taraxerone	Rauwolscine
Syringin	Hederagenin	

Adapted from Jensen, W., Fremer, K. E., Sierila, P., and Wartiovaara, V., *The Chemistry of Wood*, Browning, B. L., Ed., R. E. Krieger, Huntington, N.Y., chap. 12.

Table 23
SUGAR COMPOSITION OF BARK HYDROLYZATES

Species	% Glu	% Man	% Gal	% Xyl	% Ara	% Rha	% Ura	% Ac
Ginkgo biloba								
Inner bark	38.3	3.4	4.5	4.0	6.2	—	11.5	0.2
Outer bark	23.8	2.1	1.7	3.5	3.5	—	9.4	0.3
Pinus contora								
Inner bark	40.9	2.5	4.3	3.7	10.6	—	9.9	0.2
Outer bark	26.8	2.5	4.2	3.4	5.5	—	7.7	0.8
Pinus taeda								
Inner bark	21.3	2.5	3.1	2.1	5.6	0.3	4.6	—
Outer bark	15.8	2.6	2.5	3.8	1.8	0.1	2.1	—
Betula papyrifera								
Inner bark	28.0	0.2	1.0	21.0	2.7	—	2.2	—
Picea abies	36.6	6.5	1.3	4.8	1.8	0.3	—	—
Picea engelmannii	35.7	2.9	2.4	3.8	3.3	—	—	—
Quercus robur	32.3	0.5	1.3	16.4	2.0	0.5	—	—

Adapted from Fengel, D. and Wegener, G., *Wood: Chemistry, Ultrastructure, Reactions*, Walter de Gruyter, Berlin, 1983, chap. 9.

Table 24
PERCENT CARBOHYDRATE COMPOSITION OF VARIOUS BARKS

Carbohydrate	*Quercus robur*	*Fagus sylvatica*	*Pinus sylvatica*	*Picea abies*
Cellulose	25.4	26.2	23.1	31.7
Glucomannan	0.9	0.3	—	—
O-Acetyl-galactoglucomannan	—	—	8.4	10.1
O-Acetyl-4-*O*-methylglucuronoxylan	20.1	24.6	—	—
Arabino-4-*O*-methylglucuronoxylan	—	—	6.6	5.4
Pectin	3.8	7.4	3.6	2.5
Starch	0.6	0.4	0.2	0.6
Combined glucose	6.2	2.9	5.1	1.9

Adapted from Fengel, D. and Wegener, G., *Wood: Chemistry, Ultrastructure, Reactions*, Walter de Gruyter, Berlin, 1983, chap. 9.

Table 25
POLYOSES IDENTIFIED IN VARIOUS BARKS

Types/species	Yield %	Sugar units	Ratio	Linkage
Xylans				
Betula papyrifera	26.6	β-D-Xylp	10.0	1—4
		α-D-Me-Glu U	1.0	1—2
Populus tremuloides	18—20	β-D-Xylp	12.0	1—4
		α-D-Me-Glu U	1.0	1—2
Salix alba	3.7	Xyl	9.0	—
		Me-Glu U	1.0	—
Abies amabilis	2.1	β-D-Xylp	6.0	1—4
		α-D-Me-Glu U	1.0	1—2
		L-Araf	0.6	1—3
Piceae engelmannii	3.7	β-D-Xylp	9.0	1—4
		α-D-Me-Glu U	1.0	1—2
		L-Araf	1.4	1—3

Table 25 (continued)
POLYOSES IDENTIFIED IN VARIOUS BARKS

Types/species	Yield %	Sugar units	Ratio	Linkage
Mannans				
S. alba	1.6	Man	1.4	—
		Glu	1.0	—
S. alba	2.0	Man	1.1	—
		Glu	1.2	—
		Gal	0.5	—
P. tremuloides	Low	β-D-Man	1.3	1—4
		β-D-Glu	1.0	1—4
		α-D-Gal	0.5	1—6
Picea glauca	Low	Man	4.5	—
		Glu	1.0	—
		Gal	0.5	—
A. amabilis	3.0	β-D-Man	2.5	1—4
		β-D-Glu	1.0	1—4
		α-D-Gal	0.1	1—6
P. engelmannii	2.0	β-D-Manp	3.0	1—4
		β-D-Glup	1.0	1—4
		α-D-Galp	0.2	1—4
Pinus sylvestris	2.6	β-D-Manp	1.7	1—4
		β-D-Glup	1.0	1—4
		α-D-Galp	0.08	1—6
Glucans				
P. sylvestris	2.5	β-D-Glup	—	1—3
P. engelmannii	1—2	β-D-Glup	4.0	1—4
		β-D-Xylp	3.0	1—4
		D-Galp	1.0	1—6
Galactans				
S. alba		β-D-Gal	—	1—4
		Ara	—	1—6
P. glauca	Low	Gal	10.0	—
		Ara	1.0	—
Arabinans				
P. tremuloides		α-L-Araf	—	1—5
P. sylvestris	2.5	α-L-Araf	—	1—5
Galacturonans				
B. papyrifera	3—4	α-D-Gal	9.0	1—4
		L-Araf	1.0	—
P. glauca	7.0	Galu	—	—
		Ara	—	—
		Gal	—	—
A. amabilis	0.6	α-D-Gal U	10.0	1—4
		L-Araf	1.0	1—2,3
		D-Galp	2.0	1—2,3

Adapted from Fengel, D. and Wegener, G., *Wood: Chemistry, Ultrastructure, Reactions*, Walter de Gruyter, Berlin, 1983, chap. 9.

Table 26
CONDENSED TANNIN CONTENT OF VARIOUS BARKS

Species	Yield %
Sequoia sempervirens	2—8
Pinus densiflora	6
Robinia pseudoacacia	7
Pinus ponderosa	5—11
Larix decidua	5—20
Pseudotsuga menziesii	5—25
Picea abies	5—18
Castanea sativa	8—14
Tsuga canadensis	10—11
Betula alba	10—15
Larix leptolepis	10—25
Eucalyptus wandoo	13—15
Pinus nigra	13—25
Quercus robur	12—16
Tsuga heterophylla	15—16
Pinus sylvestris	16
Pinus radiata	17—18
Picea sitchensis	11—37
Eucalyptus adstringens	40—54

Adapted from Fengel, D. and Wegener, G., *Wood: Chemistry, Ultrastructure, Reactions,* Walter de Gruyter, Berlin, 1983, chap. 9.

Table 28
ALKALOIDS FOUND IN BARK

Alkaloid group	Individual alkaloids
Berberis and Hydrastis	Berberine
	Canadine
	Hydrastine
	Berbamine
Cinchona	Quinine
	Cinchonine
	Quinidine
	Aricine
Holarrhena	Conessine
	Kurchine
Pomegranate	Pelletierine
	Pseudopelletierine
Strychnos	Strychnine
	Brucine
Yohimbe and Quebracho	Yohimbine
	Aspidospermine
Rauwolfia	Rauwolfine
	Fauwolscine

Adaped from Jensen, W., Fremer, K. E., Sierila, P., and Wartiovaara, V., *The Chemistry of Wood,* Browning, B. L., Ed., R. E. Krieger, Huntington, N.Y., 1975, chap. 12.

Table 27
pH VALUES FOR VARIOUS BARKS AND WOODS

	pH of	
Species	Bark	Wood
Picea abies	3.6	4.6
Pinus sylvestris	3.5	4.2
Fagus sylvestris	5.0	5.3
Quercus robur	3.9	4.8

Adapted from Fengel, D. and Wegener, G., *Wood: Chemistry, Ultrastructure, Reactions,* Walter de Gruyter, Berlin, 1983, chap. 9.

Table 29
VITAMIN COMPOSITION[a] OF VARIOUS BARKS

	Vitamin			
Species	C	B_1	B_2	Nicotinic acid
Alder	15.6	0.226	0.077	10.00
Ash	8.9	0.140	0.290	2.25
Beech	5.6	0.223	0.156	2.10
Birch	2.4	0.088	0.068	0.66
Larch	13.8	0.100	0	2.00
Lime	6.8	0.145	0.160	0.90
Oak	32.0	0.074	0.110	2.15
Pine	5.0	0.055	0	2.80
Spruce	17.1	0.083	0	2.00

[a] mg/100 g bark.

Adapted from Jensen, W., Fremer, K. E., Sierila, P., and Wartiovaara, V., *The Chemistry of Wood,* Browning, B. L., Ed., R. E. Krieger, Huntington, N.Y., 1975, chap. 12.

Table 30
STRUCTURAL COMPARISONS OF HARDWOOD AND SOFTWOOD LIGNINS

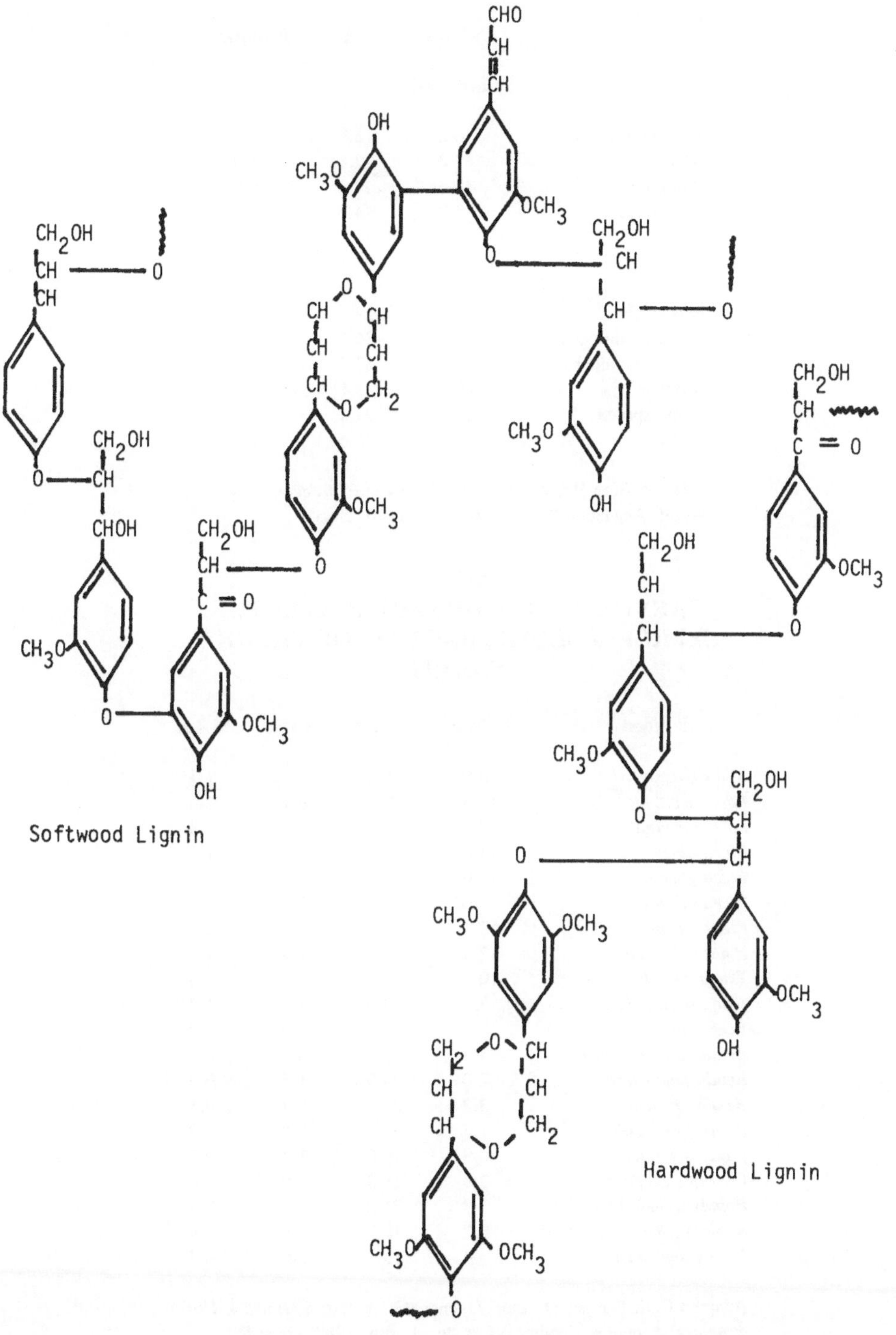

Adapted from Baltes, W., Wittkowski, R., Sochtig, I., Block, H., and Toth, L., *The Quality of Foods and Beverages*, Vol. 2, Charalambous, G. and Inglett, G., Eds., Academic Press, New York, 1981, chap. 1.

Table 31
MAJOR COMPOSITION OF HARDWOODS AND SOFTWOODS (%)

	Cellulose	Lignin	Pentosan
	Hardwood		
Trembling aspen	56.6	16.3	18.9
Beech	45.8	22.1	21.0
White birch	44.5	18.9	25.5
Red maple	44.7	24.0	20.2
	Softwood		
Balsam fir	47.8	29.4	7.7
Eastern white cedar	48.6	30.7	11.4
Eastern hemlock	45.2	32.5	7.0
Jackpine	45.0	28.6	11.4
White spruce	48.5	27.1	10.6
Tamarack	47.8	28.6	7.9

Adapted from Wenzl, H. F. J., *The Chemical Technology of Wood,* Academic Press, New York, 1970, chap. 3.

Table 32
CARBOHYDRATE DIFFERENCES IN THE HEMICELLULOSE FRACTION OF VARIOUS WOODS

Wood	% Man	% Xyl	% Gal	% Ara
Abies balsamea	10.0	5.2	1.0	1.1
Larix decidua	11.5	5.1	6.1	2.0
Larix laricina	12.3	6.0	2.4	1.3
Picea abies	13.6	5.6	2.8	1.2
Picea glauca	12.0	7.0	1.9	1.1
Picea mariana	9.4	6.0	2.0	1.5
Pinus strobus	8.1	7.0	3.8	1.7
Pinus sylvestris	12.4	7.6	1.9	1.5
Tsuga canadensis	10.6	3.3	1.8	1.0
Thuja occidentalis	7.4	3.8	1.5	1.7
Acer rubrun	3.3	18.1	1.0	1.0
Betula alleghaniensis	1.8	18.5	0.9	0.3
Betula papyrifera	2.0	23.9	1.3	0.5
Betula verrucosa	3.2	24.9	0.7	0.4
Fagus grandifolia	1.8	21.7	0.8	0.9
Fagus sylvatica	0.9	19.0	1.4	0.7
Fraxinus excelsior	2.8	18.3	0.9	0.6
Populs tremuloides	3.5	21.2	1.1	0.9
Robinia pseudoacacia	2.2	16.7	0.8	0.4
Ulmus americana	3.4	15.1	0.9	0.4

Adapted from Fengel, D. and Wegener, G., *Wood: Chemistry, Ultrastructure, Reactions*, Walter deGruyter, Berlin, 1983, chap. 5.

REFERENCES

1. **Pettersen, R. C.,** The chemical composition of wood, in *The Chemistry of Solid Wood,* Rowell, R., Ed., American Chemical Society, Washington, D.C., 1984, chap. 2.
2. **Immergut, E. H.,** Cellulose, in *The Chemistry of Wood,* Browning, B. L., Ed., R. E. Krieger, Huntington, N.Y., 1975, chap. 4.
3. **Wenzl, H. F. J.,** The chemistry of wood, in *The Chemical Technology of Wood,* Academic Press, New York, 1970, chap. 3.
4. **Fengel, D. and Grosser, D.,** Lignin structure, *Holz Roh. Werkst.,* 33(11), 32, 1975.
5. **Gardner, K. H. and Blackwell, J.,** The structure of lignin, *Biochem. Biophys. Acta,* 343, 232, 1974.
6. **Ranby, B. G. and Rydholm, A. S.,** Cellulose and cellulose derivatives, *in Polymer Processes,* Schildknecht, C., Ed., Interscience, New York, 1956, chap. 4.
7. **Schuerch, C.,** The hemicelluloses, in *The Chemistry of Wood,* Browning, B. L., Ed., R. E. Krieger, Huntington, N.Y., 1975, chap. 5.
8. **Sarkanen, K. V.,** Wood lignins, in *The Chemistry of Wood,* Browning, B. L., Ed., R. E. Krieger, Huntington, N.Y., 1975, chap. 6.
9. **Freudenberg, K.,** Lignin: its constitution and formation from *p*-hydroxycinnamyl alcohols, *Science,* 148, 595, 1965.
10. **Buchanan, M. A.,** Extraneous components of wood, in *The Chemistry of Wood,* Browning, B. L., Ed., R. E. Krieger, Huntington, N.Y., 1975, chap. 7.
11. **Jensen, W., Fremer, K. E., Sierila, P., and Wartilovaara, V.,** The chemistry of bark, in *The Chemistry of Wood,* Browning, B. L., Ed., R. E. Krieger, Huntington, N.Y., 1975, chap. 12.
12. **Gilbert, J. and Knowles, M. R.,** The chemistry of smoked foods: a review, *J. Food Technol.,* 10, 245, 1975.
13. **Baites, W., Wittkowski, R., Sochtig, I., Block, H., and Toth, L.,** Ingredients of smoke and smoke flavor, in *The Quality of Foods and Beverages,* Vol. 2, Charalambous, G. and Inglett, G., Eds., Academic Press, New York, 1981, chap. 1.
14. **Ritter, G. J. and Fleck, L. C.,** The results of analysis of heartwood and sapwood of some American woods, *Ind. Eng. Chem.,* 15, 1055, 1923.
15. **Goldstein, I. S.,** Degradation of wood by chemicals, in *The Chemistry of Solid Wood,* Rowell, R., Ed., American Chemical Society, Washington, D.C., 1984, chap. 15.
16. **Sjostrom, E.,** Bark, in *Wood Chemistry — Fundamentals and Applications,* Academic Press, New York, 1981, chap. 6.
17. **Fengel, D. and Wegener, G.,** Constituents of bark, in *Wood: Chemistry, Ultrastructure, Reactions,* Walter de Gruyter, Berlin, 1983, chap. 9.
18. **Fengel, D. and Wegener, G.,** Cellulose, in *Wood: Chemistry, Ultrastructure, Reactions,* Walter de Gruyter, Berlin, 1983, chap. 4.
19. **Fengel, D. and Wegener, G.,** Polyoses (Hemicelluloses), in *Wood: Chemistry, Ultrastructure, Reactions,* Walter de Gruyter, Berlin, 1983, chap. 5.
20. **Sjostrom, E.,** Lignin, in *Wood Chemistry — Fundamentals and Applications,* Academic Press, New York, 1981, chap. 4.
21. **Fengel, D. and Wegener, G.,** Extractives, in *Wood: Chemistry, Ultrastructure, Reactions,* Walter de Gruyter, Berlin, 1983, chap. 7.

Chapter 2

THERMAL REACTIONS OF WOOD

I. INTRODUCTION

As would be expected, the thermal decomposition of wood can be influenced by numerous factors including obviously temperature, wood composition, amount of oxygen present, and amount of water vapor available during actual pyrolysis.

Probably of the above factors, temperature is the most important, as can be seen in Table 1, the hemicellulose fraction is the first major wood fraction to undergo degradation, while the lignin fraction is the most resistant. Thus, if a relatively low temperature is used, lignin may not be completely degraded, and therefore, the resulting smoke will have a different chemical composition than if a higher temperature was used. In general, the pyrolysis of wood is fairly complete at 500°C.

Even at a temperature slightly above 100°C, significant changes can occur in wood. First, hygroscopic water is evaporated, and then bound water is released. In addition, water can be split from the hydroxyl groups of adjacent polysaccharide chains, thus forming new hydrogen bonds. At low temperatures, acetic acid also forms, while at higher temperatures (200 to 280°C), both acetic and formic acids occur, and at above approximately 280°C, the reaction can become exothermic.[2] Through the use of differential thermal analysis it has been demonstrated that during the thermal degradation of wood an endothermal maximum occurs at 120 to 150°C due to the evaporation of water. This is followed by several exothermal peaks at approximately 200 to 250°C, 280 to 320°C, and 400°C which represent the degradation of hemicellulose, cellulose, and lignin, respectively.[9]

The thermal decomposition of wood has been shown to follow an overall first-order reaction.[9] In the lower temperature range an activation energy of 105 kJ/mol has been calculated while in the higher temperature range activation energy drops to 63 kJ/mol.[9]

As would be expected, the heat of combustion for wood and its components are different. As seen in Table 2 , lignin is the major contributor and softwoods generally have higher values than hardwoods.

The presence or absence of air also has a significant effect on the thermal reactivity of wood. If air is not present, a combination of water vapor, organic liquids, and gases is released, while a mixture of tar, pitch, and charcoal remain. Charcoal, in turn, can undergo further chemical changes up to 1000°C.[3] In the presence of air, as temperature is increased, combustion results. Perhaps at this point more detail should be given to the thermal reactions associated with the three major components of wood.

II. THERMAL DECOMPOSITION OF HEMICELLULOSE

As shown in Table 1, the first major component to undergo thermal decomposition is hemicellulose. In general, hemicellulose thermally decomposes to yield furan and its derivatives, as well as a series of aliphatic carboxylic acids.[4] In addition, since hardwoods contain primarily pentosan-based hemicellulose compared to hexosan-based hemicellulose in softwoods, the latter produce more acids than softwoods upon decomposition. Thus, by combining a low temperature (less than 260°C) and using hardwood, one can generate a smoke source that is high in furans and acids, however, many furan derivatives are not stable and can undergo extensive oxidation reactions.

During the thermal degradation of hemicellulose both the acetyl and *o*-methyl groups can react to form volatile degradation products. As seen in Table 3, furan derivatives are the major intermediates.

Table 1
INFLUENCE OF TEMPERATURE ON THE THERMAL DESTRUCTION OF WOOD

Temperature (C°)	Reaction
Up to 170	Loss of water, drying
200—260	Decomposition of hemicellulose
260—310	Decomposition of cellulose
310—500	Decomposition of lignin
Above 500	Secondary reactions including oxidation, polymerization, condensation, and pyrolysis

Adapted from Baltes, W., Wittkowski, R., Sochtig, I., Block, H., and Toth, L., in *The Quality of Food and Beverages*, Vol. 2, Charalambous, G. and Inglett, G., Eds., Academic Press, New York, 1981, chap. 1.

Table 2
HEATS OF COMBUSTION FOR VARIOUS WOODS AND FRACTIONS

Source	Char (cal/g)	Gas (cal/g)	Total (cal/g)
Softwood	−1987	−3169	−5156
Hardwood	−1546	−3072	−4618
Lignin	−4375	−1995	−6370
Cellulose	−1050	−3093	−4143

Adapted from Fengel, D. and Wegener, G., in *Wood: Chemistry, Ultrastructre, Reactions*, Walter de Gruyter, Berlin, 1984, chap. 12.

III. THERMAL DECOMPOSITION OF CELLULOSE

Cellulose is the next major wood component to undergo thermal decomposition and it appears to occur by two distinct pathways, with one pathway being prevalent at lower temperatures, and the other pathway dominating at temperatures above 300°C.[6]

At temperatures below 300°C, cellulose is decomposed by the reduction in the degree of polymerization by bond scission; elimination of water; formation of free radical, carbonyl, carboxyl, and hydroperoxide groups; evolution of carbon monoxide and carbon dioxide; and production of a char residue.[6] Actual loss of cellulose polymerization can occur in the 150 to 190°C temperature range in the absence or presence of air.

Bond scission and the formation of carbon monoxide and carbon dioxide in turn occur at different rates, dependent upon the absence or presence of air. For example, carbon monoxide and carbon dioxide are produced at a faster rate in air than in its absence; however, in the presence of nitrogen, the rate of bond scission is greater than the amount of combined carbon monoxide and carbon dioxide produced.[6] The generation of carbon monoxide and carbon dioxide is due to decarbonylation and decarboxylation, respectively.

In addition, Shafizadeh[6] has proposed that the thermal degradation of cellulose may also involve a free radical mechanism. Although no free radicals have been isolated, their si-

Table 3
THERMAL DEGRADATION PRODUCTS FROM HEMICELLULOSE

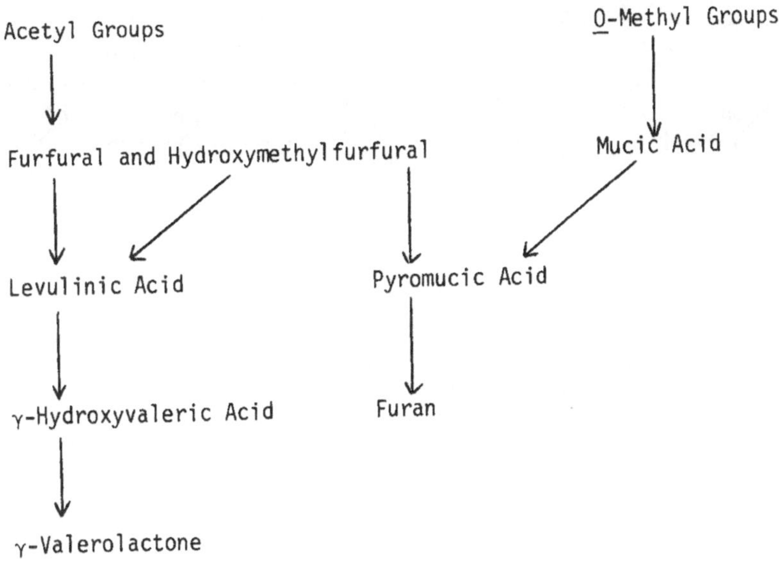

Adapted from Fengel, D. and Wegener, G., in *Wood: Chemistry, Ultrastructure, Reactions*, Walter de Gruyter, Berlin, 1984, chap. 12.

multaneous formation and decomposition have been measured, and they are thought to contribute significantly to bond scission.[6] Therefore, the first pathway of low-temperature cellulose decomposition is thought to involve three distinct steps; the initiation of pyrolysis, propagation, and product formation. The presence of oxygen or inorganic impurities accelerates the formation of free radicals, which in turn promotes bond scission, oxidation, and decomposition of cellulose to produce char, water, carbon monoxide, and carbon dioxide. The overall chemical formation and decomposition of cellulose hydroperoxide in air is summarized in Table 4. In the absence of air, a similar reaction could occur, except a lactone could be formed by rearrangement and further decomposed by dehydration and decarboxylation.[6]

The second proposed pathway, which becomes predominant as the temperature goes above 300°C, involves cleavage of molecules by transglycosylation, fission, and disproportionate reactions producing a combination of tarry anhydro sugars and lower molecular weight volatile compounds.

The initial step involves depolymerization by transglycosylation resulting in the production of levoglucosan (1,6-anhydro-β-D-glucopyranose), its furanose isomer (1,6-anhydro-β-D-glucofuranose), and numerous randomly linked oligosaccharides. In addition to intermolecular and intramolecular transglycosylations, hydration followed by fission and disproportionation reactions occurs in the gas phase, while additional decomposition and condensation reactions happen in the solid phase, thereby producing a variety of gases and volatile compounds, along with char. The pyrolysis of cellulose to the intermediate anhydro sugars is shown in Table 5. Levoglucosan is the key anhydro sugar, since yields of up to 50% have been obtained through the controlled pyrolysis of cellulose.[3] If temperature is increased further, as can be seen in Table 6, the degradation products of cellulose differ significantly. The amount of tar, reducing sugar, levoglucosan, and furanose isomer of levoglucosan

Table 4
THEORETICAL FORMATION AND DECOMPOSITION OF CELLULOSE HYDROPEROXIDE IN AIR

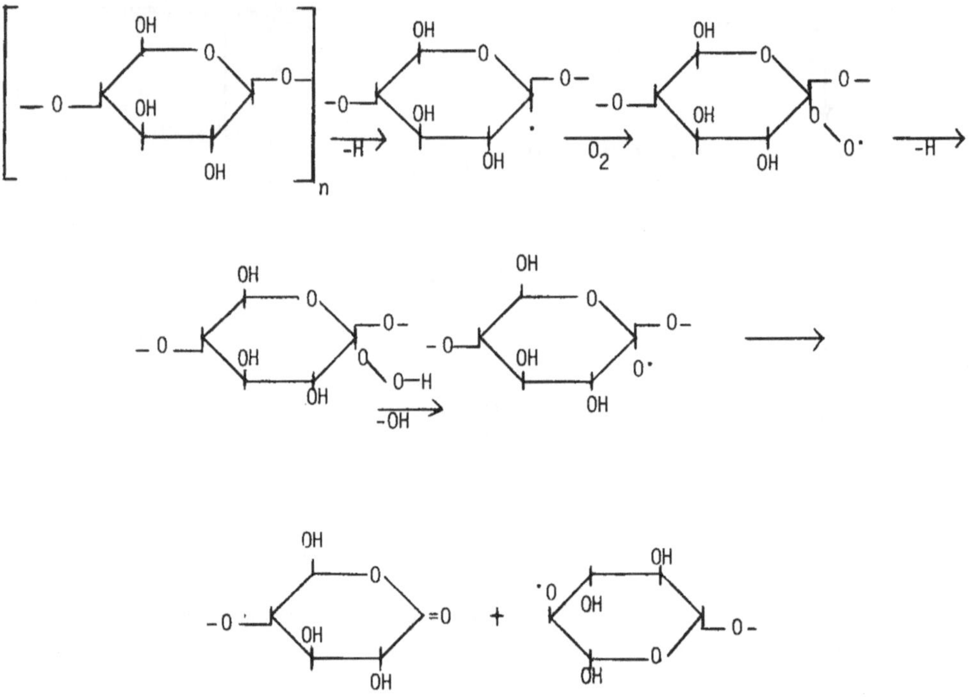

Adapted from Shafizadeh, F., in *The Chemistry of Solid Wood*, Rowell, R., Ed., American Chemical Society, Washington, D.C., 1984, chap. 13.

gradually increase, while the amount of char decreases. The tar contains anhydro sugar derivatives that can be hydrolyzed to reducing sugar. Also, the evaporation of levoglucosan and resulting volatile pyrolysis products is highly endothermic, especially at high temperatures, and thus the actual pyrolysis process may be controlled by the rate of heat transfer rather than chemical reaction kinetics.

Air velocity also can have a significant influence on the yield of various pyrolysis products associated with cellulose, since if initial volatile degradation products are not removed from the reaction site, more product decomposition can occur through interaction. As seen in Table 7, major compositional differences can result.

The composition of cellulose-derived tar has been reported[8] and compounds found include levoglucosan and 1,6-anhydro-β-D-glucofuranose as major components. Minor components include numerous compounds associated with the dehydration of glucose units. These include the compounds shown in Table 8.

Upon further heating, the end products shown in Table 9 were found. These were derived from fission or fragmentation followed by dehydration, disproportionation, decarboxylation, and decarbonylation. Russell et al.[7] recently have identified 86 compounds, including 12 phenolics, 12 aromatics, 16 aliphatic and cyclic hydrocarbons, 8 aliphatic and cyclic alcohols, 25 ketones, aldehydes, and acid esters, 13 furans, and 2 polyfunctionals associated with condensation reactions of cellulose degradation products. The overall reaction of cellulose to pyrolysis and combustion has been represented by Shafizadeh[6] as shown in Table 10.

Table 5
ANHYDRO SUGARS OBTAINED FROM THE PYROLYSIS OF CELLULOSE

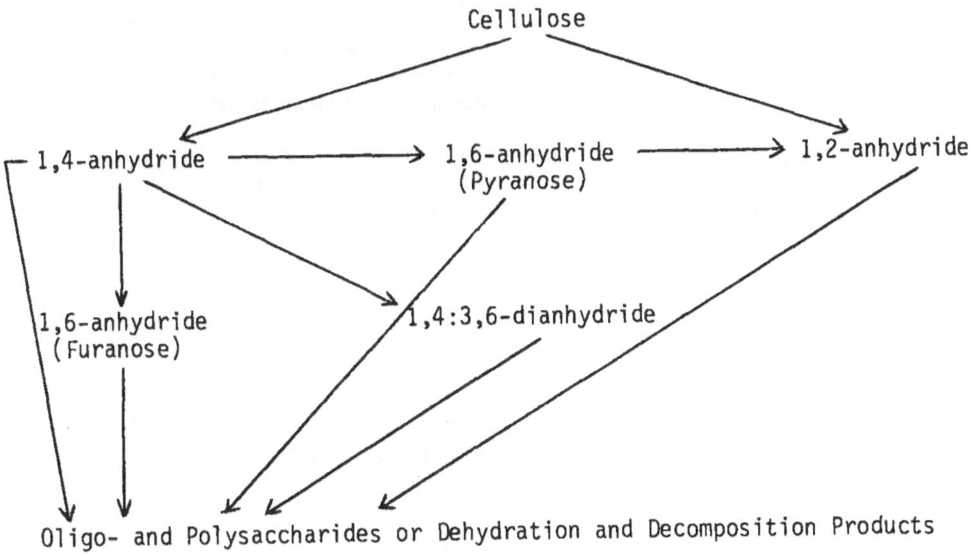

Adapted from Shafizadeh, F., in *The Chemistry of Solid Wood,* Rowell, R., Ed., American Chemical Society, Washington, D.C., 1984, chap. 13.

Table 6
INFLUENCE OF TEMPERATURE ON THE RELATIVE AMOUNTS OF CELLULOSE DEGRADATION PRODUCTS

	(%)				
Temp (°C)	**Tar**	**Reducing sugar**	**Levoglucosan**	**Furanose isomer**	**Char**
300	59	40	28	—	20
325	70	58	37	—	10
350	71	57	38	—	9
375	71	62	38	—	7
400	76	63	38	7	6
425	76	62	40	7	6
450	76	62	39	6	6
475	77	62	39	8	5
500	77	62	39	8	5

Adapted from Shafizadeh, F., in *The Chemistry of Solid Wood,* Rowell, R., Ed., American Chemical Society, Washington, D.C., 1984, chap. 13.

Earlier, the concept that cellulose can exist in various crystalline forms (I, II, III, and IV) was discussed. It has been reported[9] that the thermal stability of form III is less than that for the other forms. Also, the thermal production of levoglucosan occurs predominately in these crystalline forms as compared to the fibrous forms of cellulose which primarily degrade to furan derivatives. However, less levoglucosan results during the thermal degradation of form II as compared to the other crystalline forms.[9]

In summary, the following reactions are thought to occur during the thermal degradation of cellulose.

Table 7
INFLUENCE OF REACTION PRESSURE ON AMOUNTS OF CELLULOSE PYROLYSIS PRODUCTS

	% produced in	
Product	**Vacuum**	**Atmospheric pressure**
Char	17.8	34.2
Tar	55.8	19.1
Levoglucosan	28.1	3.57
1,6-Anhydro-β-D-glucofuranose	5.6	0.38
Hydrolyzable materials	20.9	6.08

Adapted from Shafizadeh, F., in *The Chemistry of Solid Wood*, Rowell, R., Ed., American Chemical Society, Washington, D.C., 1984, chap. 13.

Table 8
CELLULOSE TAR-BASED DEHYDRATION PRODUCTS

5-Hydroxymethyl-2-furaldehyde
2-Furaldehyde
1,6-Anhydro-3,4-dideoxy-β-D-glycerohex-3-enopyranos-2-ulose
1,5-Anhydro-4-deoxy-D-hex-1-ene-3-ulose
1,4:3,6-Dianhydro-α-D-glucopyranose
2,3-Dihydro-3,5-dihydroxy-6-methyl-4H-pyran-4-one
3,5-Dihydroxy-2-methyl-4H-pyran-4-one
1,5-Anhydro-4-deoxy-D-glycerohex-1-en-3-ulose
β-D-Glucose
3-Deoxy-D-erythro-hexosulose
O-D-Glucosyllevoglucosan

Adapted from Shafizadeh, F., in *The Chemistry of Solid Wood*, Rowell, R., Ed., American Chemical Society, Washington, D.C., 1984, chap. 13.

1. Depolymerization by transglycosylation occurs at around 300°C resulting in a mixture of levoglucosan, other monosaccharide derivatives, and variously randomly linked oligosaccharides. The above is commonly called tar.
2. In addition to the above, dehydration of sugar units occurs giving a variety of unsaturated compounds as well as various furans.
3. As temperature increases, the fission of sugar units generates numerous carbonyl compounds.
4. Condensation of unsaturated products and the subsequent cleavage of the side chains leaves a reactive carbonaceous residue.

IV. THERMAL DECOMPOSITION OF LIGNIN

From a smoke flavor standpoint, the thermal destruction of lignin is quite important since some of the derived compounds include phenols and phenolic esters such as guaiacol (2-methoxyphenol) and syringol (2,6-dimethoxyphenol), along with their homologs and derivatives. A variety of associated compounds composed of methyl, ethyl, propyl, vinyl, allyl, and propenyl sidechains also result. In general, these sidechains are three or less carbon atoms in length and usually occur in the *para* position to the phenolic hydroxyl group.[3,4]

Initial decomposition of lignin occurs by fission at the heterocyclic furan, pyran rings, and ether linkages. This in turn produces guaiacol, which can be further degraded to form phenol and cresols.[3]

Table 9
PYROLYSIS PRODUCTS (600°C) OF CELLULOSE

Compound	Relative %
Acetaldehyde	2.3
Furan	1.6
Acetone/propionaldehyde	1.5
Propenal	3.2
Methanol	2.1
2,3-Butanedione	2.0
1-Hydroxy-2-propanone	2.1
Glyoxal	2.2
Acetic acid	6.7
2-Furaldehyde	1.1
Formic acid	0.9
5-Methyl-2-furaldehyde	0.7
2-Furfuryl alcohol	0.5
Carbon dioxide	12.0
Water	18.0
Char	15.0
Tar	28.0

Adapted from Shafizadeh, F., in *The Chemistry of Solid Wood*, Rowell, R., Ed., American Chemical Society, Washington, D.C., 1984, chap. 13.

Table 10
THE PYROLYSIS AND COMBUSTION OF CELLULOSE

Adapted from Shafizadeh, F., in *The Chemistry of Solid Wood*, Rowell, R., Ed., American Chemical Society, Washington, D.C., 1984, chap. 13.

In the case of hardwoods which have higher amounts of methoxy substituents, the presence of dimethoxy rings can result in the production of syringol and a whole series of *para*-substituted compounds, in addition to guaiacols.[4]

The release of ferulic acid is a key intermediate in the thermal degradation of lignin, since it can undergo further decarboxylation to yield the series of compounds seen in Table 11. However, the oxygenated compounds vanillin, vanillic acid, and acetovanillone were only found when the reaction occurred in the presence of air. Therefore, it would appear that dependent on the amount of air present during the pyrolysis of lignin, the end product composition can be controlled.

Fenner and Lephardt[8] followed lignin decomposition over a wide temperature range and

Table 11
THERMAL DEGRADATION PRODUCTS OF FERULIC ACID

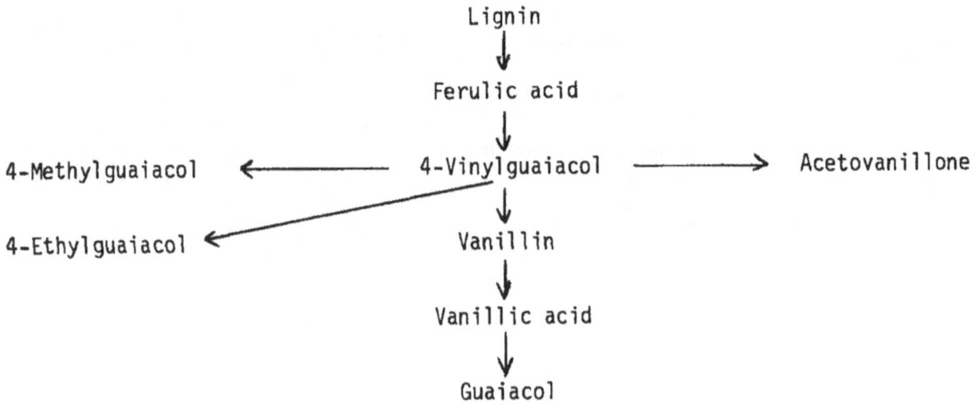

Adapted from Gilbert, J. and Knowles, M. E., *J. Food Technol.*, 10, 245, 1975.

found that initial lignin decomposition occurred at 120 to 300°C. Compounds associated with this stage included formic acid, formaldehyde, carbon dioxide, and water, primarily resulting from single-bond fragmentation in the phenylpropane sidechains.

The major decomposition of lignin extended over the temperature range of 300 to 480°C and was maximum at 385°C. Compounds observed in this stage included methanol, carbon dioxide, carbon monoxide, water, methane, guaiacol, 2-methoxy-4-alkyl-substituted phenol, acetone, and acetic acid. They postulated that methanol was derived directly from the methoxyl substituents of lignin, whereas the other compounds were derived as fragmentation products of major chain linkages releasing monomeric phenol units into the vapor phase which undergo further secondary degradation. Maximum phenolic compound formation was observed at approximately 320°C, while at higher temperatures secondary degradation products began to appear, thus suggesting that temperatures above 320°C should not be used if one is trying to maximize phenolic yield in smoke, even though lignin will further decompose to temperatures approaching 500°C.

The thermal degradation products of wood can be grouped into the combustible volatiles, tar, and char. Tar, upon further degradation, can be converted to volatiles and char.

V. COMBUSTION

In addition, Shafizadeh[6] has shown that the char remaining after lignin pyrolysis is highly reactive and chemically different, dependent upon pyrolysis temperature, and can undergo smoldering or glowing combustion, or if the temperature is high enough and air conditions are adequate, flaming combustion can occur.

Since lignin is the primary constituent contributing to char formation, while cellulose and hemicellulose primarily contribute to the formation of volatile pyrolysis products that are responsible for flaming combustion, the higher the lignin content of wood, the higher its flaming combustion temperature requirement.

As would be expected, the higher the combustion temperature, the higher the amount of gases produced. As seen in Table 12, tar content is not significantly influenced by temperature but other components are altered.

Table 12
INFLUENCE OF COMBUSTION TEMPERATURE ON PRODUCT YIELDS

	Temp (°C)			
	540	**650**	**760**	**870**
Gases (% yield)	25	28	32	34
Char (% yield)	33	31	29	29
Water (% yield)	27	25	23	22
Tar (% yield)	16	16	19	16

Adapted from Fengel, D. and Wegener, G., in *Wood: Chemistry, Ultrastructure, Reactions,* Walter de Gruyter, Berlin, 1984, chap. 12.

REFERENCES

1. **Baltes, W., Wittkowski, R., Sochtig, I., Block, H., and Toth, L.,** Ingredients of smoke and smoke flavor, in *The Quality of Food and Beverages,* Vol. 2, Charalambous, G. and Inglett, G., Eds., Academic Press, New York, 1981, chap. 1.
2. **Wistreich, H. E.,** Smoking of meats, Proc. Meat Ind. Res. Conf., Chicago, 1977, 37.
3. **Goos, A. W.,** The thermal decomposition of wood, in *Wood Chemistry,* Vol. 2, Wise, L. E. and Jahn, E. C., Eds., Reinhold, New York, 1952, chap. 20.
4. **Gilbert, J. and Knowles, M. E.,** The chemistry of smoked foods: a review, *J. Food Technol.,* 10, 245, 1975.
5. **Browning, B. L.,** Composition and chemical reactions of wood, in *The Chemistry of Wood,* Browning, B. L., Ed., R. E. Krieger, Huntington, N.Y., 1975, chap. 3.
6. **Shafizadeh, F.,** The chemistry of pyrolysis and combustion, in *The Chemistry of Solid Wood,* Rowell, R., Ed., American Chemical Society, Washington, D.C., 1984, chap. 13.
7. **Russell, J. A., Miller, R. K., and Molton, P. M.,** Formation of aromatic compounds from condensation reactions of cellulose degradation products, *Biomass,* 3, 43, 1983.
8. **Fenner, R. A. and Lephardt, J. O.,** Examination of the thermal decomposition of kraft pine lignin by fourier transform infrared evolved gas analysis, *J. Agric. Food Chem.,* 29, 846, 1981.
9. **Fengel, D. and Wegener, G.,** Influence of temperature, in *Wood: Chemistry, Ultrastructure, Reactions,* Walter de Gruyter, Berlin, 1984, chap. 12.

Chapter 3

SMOKE GENERATION TECHNOLOGY

I. INTRODUCTION

Through the years various ways of pyrolyzing wood to produce smoke-flavor compounds have developed. From a commercial standpoint this has evolved from the first primitive ways which probably just involved hanging meat on the ceiling of a dwelling and permitting smoke resulting from a fire that was used solely as a source of warmth to pass around and through the product. Today, the generation of smoke in many commercial operations can be quite sophisticated. Also, in most countries techniques are required that minimize environmental pollution.

Another factor that in some countries has dictated that technologically advanced smoke-generation systems be employed is the potential health concerns associated with certain rather primitive smoke-generation systems. This potential problem will be discussed in a later chapter.

The current major commercial methods of generating smoke from wood have been reviewed,[1-3] and thus at this point, only a brief summation of the six major techniques will be given. It should be noted that modern smoking operations usually pyrolyze wood in a separate chamber and then transport the generated smoke into a chamber where the food item is present. This is in contrast with more primitive techniques where smoke generation and application steps are accomplished in the same chamber.

In fact, some of the more advanced smoke-generation units are photoelectronically controlled in such a manner that the appearance of a flame in the smoldering sawdust causes the metering of a new portion of sawdust. As a result, such units can efficiently produce a smoke of uniform composition.[55]

Other researchers[56] have thoroughly studied the aerodynamic parameters that are important for the optimum mixing of air and smoke so as to obtain maximum smoke color and flavor in the food being smoked as well as to minimize smoke pollution. This information has gone into the design of smoke-processing chambers. Also, it has been found that pulsating air is more effective than a steady-flow air system.[56] In addition, computer simulations of many smoking processes have been developed[57] to predict the most efficient and economic options required to produce and utilize smoke in a wide variety of foods.

II. SMOLDERING

As the name smoldering implies, a wood source, usually in a sawdust form, is permitted to smolder by the action of a heating coil, or in some instance, in the presence of a gas flame. Through continuous gentle agitation, and in the presence of sufficient oxygen, a steady supply of sawdust and resulting smoke will be produced.

The resulting smoke is then either drawn or pushed into the actual smoking chamber and permitted to interact with the food in question. The amount of tar and other solid particles that enters the smoking chamber can be controlled to some extent by the distance the smoldering unit is from the smoking unit, with the longer the distance, the greater the opportunity for tar to settle out. Also, various baffles and filters can be placed in line to remove tar, but routine cleaning is required to remove buildups of these compounds in the system.

Smoke production or generation temperature, using the smoldering technique, can vary over a fairly wide range, dependent upon available air. If a large quantity of air is present, the actual smoldering temperature may be in excess of 800°C, but if the amount of available air is controlled, lower smoldering temperatures will result.

Smoke production temperature can also be lowered by controlling the sawdust moisture content.[2] Most small commercial operations do not normally consider wood moisture content, but wood containing approximately 30% moisture will effectively lower smoldering temperature 100 to 300° over a completely dry wood sample. This in turn reduces energy requirements and produces a smoke generally lower in polycyclic aromatic hydrocarbons (PAH), which have significant health implications.

III. FRICTION

Some processors use the friction method whereby an intact wood block is pressed against a rapidly rotating metal friction wheel. Heat produced by friction is sufficient to cause wood pyrolysis. Air is permitted to enter through the center of the friction wheel thus producing a cooling effect on the resulting smoke. The more air that is present, the lower the resulting smoke temperature. This is in direct contrast with the smoldering technique. In addition, it is a flameless method of generating smoke and the resulting wood dust falls to the bottom of the generation chamber with the smoke being directed into the smoking chamber.

If the apparatus is operated in a discontinuous fashion, such that the friction wheel is in operation for 10 sec and then rested for several minutes, smoke temperatures that are only a few degrees above ambient temperature can be achieved. However, if operated in a continuous fashion, smoke temperatures in the range of 140 to 160°C are common.

Other factors influencing smoke temperature include the amount of force applied to the wood block, the wheel speed, and the shape of the friction edges.[4-6] The actual temperature during wood pyrolysis using friction is in the neighborhood of 450 to 560°C, but rapid cooling occurs, so actual smoke temperature can be relatively low.[6]

However, Toth[4] has reported that the sensory properties resulting from friction-generated smoke, when compared to smoldering-generated smoke, contain only low amounts of volatile phenols, guaiacol, and syringol, but higher amounts of allyl- and propenyl syringol, and only about 10% of higher molecular weight compounds, which in turn adversely influence product taste and smell. Apparently, the rapid cooling associated with friction smoke generation does not permit sufficient time and/or temperature for the primary products of wood pyrolysis to undergo extensive secondary reactions, thus limiting the number of compounds formed.

IV. WET SMOKE

In the wet smoke or condensate smoke method, sawdust and superheated steam containing varying amounts of heated air are blended, causing sawdust pyrolysis. The resulting smoke has a temperature of approximately 80°C, and is moist because of condensed water from steam. This represents another flameless method, and pyrolysis normally occurs in the 300 to 400°C range. Advantages cited for the process include a short treatment time and improved smoke yield.[7-11]

V. FLUIDIZATION

With this technique, air is heated to between 300 to 400°C by means of an electric heater, and under high velocity, mixed with sawdust. Because of the resulting turbulence, the sawdust is "suspended" or fluidized. Because of the air temperature, pyrolysis occurs at approximately 350°C in around 10 sec. A cyclonic separator is used to yield smoke and charred sawdust. This procedure is also flameless, and because of the cooling effect, smoke temperature does not go much beyond ambient temperature.

Table 1
SMOKE CHARACTERISTICS OF VARIOUS SMOKE PRODUCTION TECHNIQUES

Smoke production technique	Smoke production temp (°C)	Smoke temp (°C)	Smoke characteristics
Smoldering	400—800	20	Dry, dense, contains oxygen
Friction	300—500	20	Dry, dense, much oxygen
Wet smoke	300—400	80	Moist, dense, low oxygen
Fluidization	300—400	25	Dry, dense, contains oxygen
Two stage	300—400	30	Dry, dense, contains oxygen
Carbonization	300—400	20	Dry, dense, low oxygen

VI. TWO STAGE

The two-stage procedure is a modification and extension of the fluidized technique described above. The first stage involves initial wood pyrolysis by the action of nitrogen or carbon dioxide at a temperature of 300 to 400°C on sawdust. In the second stage, the reactants of the first stage are mixed with oxygen or air heated to 200°C, and further pyrolysis occurs. The second stage promotes oxidation, condensation, and polymerization reactions thus resulting in a more complex mixture of potential flavor compounds in the smoke. Because of the dual effect, the resulting temperature is in the neighborhood of 30°C.

Factors influencing smoke composition and yield are first-stage pyrolysis temperature, amount of nitrogen or carbon dioxide during first-stage pyrolysis, and temperature of second-stage oxidation. Maximum smoke amounts were found to occur at a first-stage temperature of 400°C, with a nitrogen level of 1500 ℓ/hour, and a second oxidation temperature of 200°C.[2]

VII. CARBONIZATION

Production of smoke via carbonization requires specialized equipment in that sawdust is pressed together in a tubular casing by means of a tapered screw. As a result, most of the air associated with sawdust is eliminated. Directly at the end of the compression device is a variable temperature electrical heating element, which, because of lack of air in the compressed sawdust, causes the product to give off smoke during carbonization. This is another system that is flameless, and produces smoke that has a temperature close to ambient conditions.

Some of the general characteristics of the smoke produced by the above six techniques are summarized in Table 1. It should be noted that with the smoldering and friction methods, smoke-generation temperature is rather difficult to control, whereas with the other procedures, it is easier to generate smoke at a preselected temperature, which is usually lower than that used for the first two mentioned methods.

Also, with the smoldering methods, flames may actually be produced, which can actually combust desirable smoke-associated compounds. The other methods in turn all produce smoke without the presence of flames.

You can also note from Table 1 that only the wet smoke method results in moist smoke, while the others produce dry smoke. This in turn could influence product yield due to potentially higher food weight losses associated with the use of dry smoke. Also, a moist smoke may be more beneficial from a color standpoint.

The other major difference among the procedures outlined in Table 1 is the fact that the smokes produced have varying amounts of oxygen associated with them. This has both positive and negative aspects. Oxygen will provide for more secondary oxidation reactions to occur, thus resulting in the formation of additional flavor-associated compounds. However, certain smoke-associated compounds are sensitive to oxygen and are actually decomposed in the presence of large amounts of oxygen.

VIII. HOT VS. COLD SMOKING

The actual smoke temperature can vary quite widely dependent upon the method of generation and the ambient temperature surrounding the food item being smoked. Obviously, if high ambient temperatures are used, the food will also undergo some degree of cooking as well as absorb and react with components in smoke. On the other hand, it is possible to expose food to temperatures only slightly above room temperature and the food will still obtain a smoky character without undergoing any cooking. This is the major difference between cold smoking, which usually occurs at temperatures of 15 to 25°C,[29] while for hot smoking, temperatures in the range of 55 to 80°C are employed.[1] Foods typically cold smoked are raw ham and bacon, while typical hot smoke foods include cooked ham and frankfurters. Some processors also utilize warm smoking where the smoke and chamber temperature is between 25 and 40°C.[13]

Several reports have appeared demonstrating that hot vs. cold smoking can influence resulting phenol composition[14,15] with apparently more variation in the fat than in the lean portion of meat.[15] However, no difference due to hot or cold smoke processing relative to carbonyl content was found.[15]

From an efficiency standpoint, hot smoke becomes associated with food approximately seven times faster than cold smoke,[1] but with hot smoking, chemical changes occur primarily in the fat portion of the food. Essentially, a cold-smoked food is not cooked.

Time of smoking is another major difference between the two processes. For example, the cold smoking of a food can take up to several weeks so that the proper amount of color and/or flavor are generated in the food, whereas hot smoking may only take several hours.[30]

Cold smoking seems to be especially adaptable to most fish species.[58-60] In the case of rainbow trout[58] it was found that cold smoking for 2 hr at 30°C gave a product of equal smoke quality to that smoked for 1 hr at 80°C. With catfish and kipper, cold smoking was found to be preferable from a sensory standpoint, and in addition, better yields were obtained as compared to hot smoking of either whole fish or fillets.[59]

Some processors also practice what is called moist smoking.[12] This involves using a smoke-chamber temperature of 24 to 48°C and relative humidities in excess of 30%. The process has found application in the smoking of dry sausage and salamis.

Dark smoking is another process whereby a dense smoke is permitted to blacken the surface of the food in question.[31] As a result, the food has a strong smoky character. Smoke that has quite a bit of particulate matter associated with it, such as pine, is normally used, but the practice has to be questioned from a health standpoint, since smoke particulate matter can be high in PAH.

As can be seen in Table 2, there is no question that the composition of light and dark smoke is significantly different, thus resulting in different sensory properties. It is also apparent from Table 2 that the wood source can also influence smoke composition, and in turn its sensory properties. This concept will be expanded upon in a later chapter.

IX. ELECTROSTATIC SMOKING

As mentioned earlier, smoke is composed of two phases, namely a particulate or dispersed phase and a gaseous or dispersing phase.[32] The dispersed phase is the visable portion of smoke and is primarily composed of tars, wood resins, high-boiling phenolic-type com-

Table 2
INFLUENCE OF WOOD SOURCE AND DEGREE OF SMOKING ON RESULTING COMPOSITION

Smoke fraction	Beech smoke (light)	Beech smoke (heavy)	Oak smoke (light)	Oak smoke (heavy)
Ash and soot	8.0	8.2	21.8	23.5
Insoluble in ether	216.0	99.6	154.2	757.4
Free bases	45.3	155.8	42.0	48.0
Aldehydes and ketones	47.9	108.9	24.8	71.0
Carboxylic acids	37.3	203.6	64.3	327.5
Phenols	15.9	93.3	9.0	73.3
Neutral organic compounds	298.9	1757.0	477.6	803.6
Unknown	12.2	95.6	21.9	45.2

Adapted from Ziemba, A., *Przem. Spozyw.*, 2, 200, 1957.

Table 3
VOLATILE COMPOSITION OF UNTREATED AND ELECTROSTATICALLY FILTERED SMOKE

	mg%	
Compound class	Untreated smoke	Electrostatically filtered
Acids	211.0	150.0
Phenols	7.5	2.7
Carbonyls	347.0	192.0

pounds, and lower boiling compounds that vary with temperature and smoke concentration.[1] These particles are usually in the form of liquid droplets that form by condensation and are 0.196 to 0.346 μm in average diameter.[17-19,33] However, most of the characteristic flavor associated with wood smoke is present in the gas phase.[20]

The fact that wood smoke is a two-phase system has brought about the introduction of electrostatic smoking.[1,2,20-26,34,35,39] The procedure has found application in a continuous tunnel-type arrangement whereby the product is positioned between live electrical wires that are charged to between 20 to 60 kV. Smoke passing through this system is charged according to its phase and smoke components can then precipitate on the oppositely charged meat surface. This is usually followed by infrared radiation, which drys the product surface.[2,27]

As would be expected, the process changes the volatile composition of smoke. As seen in Table 3, electrostatically filtered smoke was found to be lower in acids, phenols, and carbonyls than untreated smoke.

Because of these differences one would predict that products smoked by the two procedures would have different sensory properties. Rusz[23] found that for a variety of smoked sausages, electrostatically filtered smoke usually gave less smoky aroma and color development than normal smoke but the sensory flavor scores were fairly similar. However, since most researchers used subjective sensory panels to evaluate the original properties of smoked foods, no clearcut decision can be made on whether or not electrostatically smoked foods are more or less acceptable than traditionally smoked foods.

Factors such as distance between wires and time of treatment are obviously two factors that can significantly influence the amount of flavor compounds deposited. For example, as seen in Table 4, wire spacing of 60 cm produced less total phenols than a traditionally smoked bacon control, while a spacing of 20 cm produced more. Thus, by varying wire spacings, one could at least theoretically obtain comparable phenol levels. Also, as seen in

Table 4
INFLUENCE OF ELECTROSTATIC PROCESSING CONDITIONS OF THE VOLATILE COMPOSITON OF BACON

Variable	Phenols[a]	Monocarbonyls[b]	Total carbonyls[b]
Traditionally smoked	3881	3.3	22
Electrostatic-20 cm wire spacing	5481	4.4	31
Electrostatic-60 cm wire spacing	3660	3.7	22
Traditionally smoked (2 hr)	3730	NR	15
Electrostatic (2 hr)	4826	NR	17
Electrostatic (1 hr)	2418	NR	10

Note: NR = not reported.

[a] mg/100 g bacon.
[b] m*M*/100 g bacon.

Adapted from Girard, J. P., Talon, R., Sirami, J., and Denoyer, C., *Sci. Aliment.*, 2, 1, 1982.

Table 4, time is influential. The 2 hr electrostatic process resulted in more phenols than the 2 hr traditional control. Therefore, an advantage of electrostatic smoking is that it is faster.

In the case of fish, it has been reported[61] that the conditions of electrostatic smoking have to be adjusted dependent upon the size and fish species being smoked. From 50 to 80% of the smoke components were found to be trapped by the skin and subcutaneous muscle. Other factors found to be important were skin density, the presence of scales, and the amount of subcutaneous fat.[62]

X. MISCELLANEOUS TECHNOLOGIES

A. Gas-Phase Smoke

Another interesting technological concept is based on utilization of only the gaseous phase of wood smoke. Generated smoke is separated by cooling and washing or filtering to recover only the gaseous phase. This usually results in a characteristic flavor but a lighter surface color can be expected, since no particulate matter is present.

B. Closed Smoking

Primarily because of concerns for environmental pollution, the closed smoking system has been developed.[2,36] In this procedure, a portion of the smoke remaining in the smoking chamber as exhaust is recycled back to the smoke-generation chamber as a partial replacement for fresh air.[28,37] This in turn has several effects. First, less total smoke is emitted into the atmosphere since it is essentially recycled. Second, since the recyled smoke is generally lower in oxygen than normal air, the smoldering temperature in the generation chamber is reduced, which in turn lowers the formation of PAH.

C. Indirect Smoking

In this process, herbs and/or spices that are going to be incorporated into a food formulation are smoked independently[45] and then just added to food, which is then usually heat processed. Obvious limitations include the fact that only low overall smoke flavor may result in the food, and smoke flavor can be localized to the spice portion of the food.

D. Liquid Smoke

In certain countries the product called liquid smoke has become quite popular. The reported

Table 5
COMPOSITION OF COMMERCIAL LIQUID SMOKES

Fraction	%
Water	11—92
Phenols	0.2—2.9
Acids	2.8—9.5
Carbonyls	2.6—4.6
Tar	1—17

Adapted from Baltes, W., Wittkowski, R., Sochtig, I., Block, H., and Toth, L., in *The Quality of Foods and Beverages*, Vol. 2, Charalambous, G. and Inglett, G., Eds., Academic Press, New York, 1981, chap. 1.

advantages of utilizing liquid smoke condensate include: (1) some flavor can be incorporated throughout a product in a uniform manner instead of having a higher concentration on the outside vs. inside as with traditional smoking, (2) it can intensify the flavor of traditionally smoked foods, (3) there is closer control on the amount of smoke flavor a product receives, (4) it can be fractionated to intensify smoke flavor and to remove potentially harmful compounds before it is used with food, (5) it has application to a wide variety of foods that traditionally are not smoked, (6) it can be used at the consumer level as well as the commercial processing level, (7) normally it represents a cost savings since wood and smoking equipment are not required as part of a food smoking plant, (8) there is less environmental pollution associated with its use, and (9) it can be applied in various ways such as spraying on the surface, dipping, and actual mixing in with the food.[1,38,46-51]

Condensed smoke can be obtained and utilized in various forms. The classical liquid smoke is smoke condensate that is dissolved in water, oil, or smoke extracts in organic solvents. In addition, smoke condensate can be adsorbed on the solids such as spices, salt, sugars, starch, or protein, thus resulting in dry or powdered forms.[64]

Freshly generated smoke that is condensed in water usually has a bright yellow color that with time darkens due to the formation of brown-colored condensation or polymerization products. These compounds, along with tarry products and polycyclic aromatic hydrocarbons (PAH), settle out with time. The resulting liquid phase can then be decanted, resulting in a liquid preparation that is low in PAH. In contrast, smoke that is condensed into water that contains solvents such as ethanol or glycol does not precipitate tars and PAH.[64]

Also, as seen in Table 5, the composition of commercial liquid smoke can vary widely, dependent upon method of manufacture.[41-44] An important compositional factor is the acidity in the resulting smoke, since this in turn can influence phenol solubility. As shown in Table 6, the solubility of most phenols increases significantly with increasing acidity. Thus, if the acidity is not high enough, some phenols may actually settle out during the prolonged storage of liquid smoke, and thus their measurement is suggested.[52]

In the case of experimentally smoked fish in a traditional manner as compared to the use of liquid smoke, it has been reported[63] that the liquid smoke treated product has less pronounced aroma and flavor. Analysis of the phenolic substances showed that the liquid smoked fish had 1.14 to 1.75 mg/100 g while the traditional product had 5.94 to 13.55 mg/100 g. Also, carbonyl and total acid levels were 22.8 and 0.59 mg/100 g, respectively, in the liquid smoke product while corresponding values for traditionally smoked fish were 23.7 and 0.69.

Table 6
INFLUENCE OF LIQUID SMOKE ACIDITY ON PHENOL SOLUBILITY

	Solubility of (%)		
Acidity (%)	**Cresol**	**Guaiacol**	**2,6-Dimethoxyphenol**
2	2	40	30
4	8	45	35
6	17	62	42
8	22	78	78
10	30	80	100

Adapted from Gubler, D. and Vernois, M., *Ind. Aliment. Agric.*, 89, 621, 1972.

The question may arise as to whether liquid smoke or traditional smoking was used in the production of certain smoked foods. A somewhat simple answer to this question is the measurement of a specific smoke component, usually guaiacol, in the center vs. the exterior of the food.[40,53,54] In traditional smoking one would expect a much lower concentration of quaiacol towards the center since phenol migration is not significant in traditional smoking. However, if liquid smoke had been added to the product, the guaiacol level should be fairly uniform throughout the product. Care should be taken though, since if liquid smoke were only sprayed on the product surface and then the product heated and/ or smoked, low guaiacol levels would also be present in the product center leading someone to falsely conclude the product was smoked in the traditional manner.

REFERENCES

1. **Draudt, H. N.,** The meat smoking process: a review, *Food Technol.*, 17, 1557, 1963.
2. **Klettner, P. G.,** Modern methods of smoking meat products, *Fleischwirtschaft*, 59(1), 73, 1979.
3. **Klettner, P. G.,** Technology of common smoking methods, *Fleischerei*, 29(10), 60, 1978.
4. **Toth, L.,** Influence of smokehouse technology on phenols in the smoke, *Fleischwirtschaft*, 60(8), 1472, 1980.
5. **Klettner, P. G.,** Curing smoke production with the friction smoke generator, *Fleischwirtschaft*, 55(11), 1498, 1975.
6. **Khavan, E. A., Blejus, V. G., Sandosvskii, B. F., and Voskresenskii, H. A.,** Smoke, *Rybn. Khoz.*, 47(9), 57, 1971.
7. **Zweig, K. and Schossig, H.,** Experience with the type 301 steam-type smoke generator in Bruhwurst sausage manufacture, *Fleisch*, 33(9), 173, 1979.
8. **Schuldt, P.,** Application of steam-generated smoke in cold-smoking and controlled environment smoking, *Fleischerei*, 30(12), 990, 1979.
9. **Fessman, G. and Fessman, K. D.,** Steam-generated smoke for hot and cold smoking, *Fleischerei*, 30(12), 1006, 1979.
10. **Rusz, J. and Kopslova, M.,** New method for smoking of meat and meat products, *Fleisch*, 28(12), 236, 1974.
11. **Fessman, G.,** Production and utilization of curing smoke, *Fleischerei*, 24(11), 14, 1973.
12. **Reich, W.,** Applications of modern universal cooking chambers for smoking and cooking, *Fleischerei*, 28(3), 25, 1977.
13. **Butanski, I.,** Smoke and its preserving role in meat and meat product smoking, *Khranit. Prom.*, 28(5), 8, 1979.
14. **Toth, L.,** Uptake of phenols from smokehouse smoke into meat products, *Fleischwirtschaft*, 62(11), 1398, 1982.

15. **Khlamova, R. I., Soloveva, L. I., and Petrakova, A. N.,** Content of phenolic and carbonyl compounds in cold smoked products, *Myasn. Ind. SSSR,* No. 2, 37, 1980.
16. **Ziemba, A.,** Generation of industrial curing smoke and its chemical composition, *Przem. Spozy.,* 2, 200, 1957.
17. **Foster, W. W.,** Deposition of unipolar charged aerosol particles by mutal repulsion, *Br. J. Appl. Phys.,* 10, 206, 1959.
18. **Foster, W. W.,** Attenuation of light by wood smoke, *Br. J. Appl. Phys.,* 10, 416, 1959.
19. **Foster, W. W., Simpson, T. H., and Campbell, D.,** Studies of the smoking process for foods. II. The role of smoke particles, *J. Sci. Food Agric.,* 12, 635, 1961.
20. **Foster, W. W. and Simpson, T. H.,** Studies of the smoking process for foods. I. The importance of vapours, *J. Sci. Food Agric.,* 12, 363, 1961.
21. **Hanley, J. W., Draudt, H. N., and Brockmann, M. C.,** A continuous process for smoked meat. I. Development of a process, *Food Technol.,* 9, 591, 1955.
22. **Hanley, J. W., Montgomery, G. I., Rarick, M. S., and Brockmann, M. C.,** A continuous process for smoked meats. II. Equipment design and application, *Food Technol.,* 9, 597, 1955.
23. **Rusz, J.,** Experiences with the use of electrostatically filtered smoke for smoking of meat products, *Acta Aliment. Pol.,* 3, 301, 1977.
24. **Girard, J. P., Talon, R., Sirami, J., and Denoyer, C.,** Electrostatic smoking: efficiency in relation to two technological parameters, *Sci. Aliment.,* 2, 1, 1982.
25. **Talon, R. and Girard, J. P.,** Smoking of meat products, *Viandes Produits Carnes,* 1(4), 16, 1980.
26. **Kleitz, R.,** Smoking equipment used in the fish industry of the German Democratic Republic, and possibilities for its mechanization, *Lebensm. Ind.,* 26, 257, 1979.
27. **Pigott, G. M. and Tillisy, M.,** Short time high quality (STHQ) smoking and dehydration, *Proc. 6th Int. Conf. Food Sci. Technol.,* 1, 193, 1983.
28. **Huch, R.,** Means of restricting emissions from plants for smoked foods, *Wasser Luft Betr.,* 20(12), 669, 1976.
29. **Corretti, K.,** Raw sausage and meat products. II. Raw meat products. Smoking and storage of uncooked cured meat, *Fleischwirtschaft,* 55(11), 1504, 1975.
30. **Toth, L.,** Influence of smoking technology on smoke components in meat products, *Fleischwirtschaft,* 62(2), 172, 1982.
31. **Shaposhnikov, Y. K., Khvan, E. A., Stepanova, M. I., and Kondakova, L. V.,** The chemical composition of a smoking medium, *Rybn. Khoz.,* No. 9, 75, 1972.
32. **Proskura, Y. D.,** Conditioning of smoking smoke, *Rybn Khoz.,* 47(10), 55, 1971.
33. **Rusanov, V. V., Lapshin, I. I., and Kuchumov, A. M.,** Study of smoking medium particles by electron microscopy, *Rybn. Khoz.,* No. 5, 67, 1975.
34. **Hardy, R. and McGill, A. S.,** Smoking of foods: methods and some toxicological aspects, *Proc. Biochem.,* 14(11), 2, 1979.
35. **Khan, E. A.,** Physico-chemical properties of wood smoke, *Tr. Vses. Nauchno-Issled. Inst. Morsk. Rybn. Khoz. Okeanogr.,* 73, 102, 1970.
36. **Abaseev, V. K., Kvasov, A. A., and Shaposhnikov, Y. K.,** Catalytic additional combustion of residual smoke gases from fish smoking, *Rybn. Khoz.,* No. 11, 66, 1973.
37. **Sadovskii, B. F., Khvan, E. A., Radakova, T. N., and Bazarov, A. P.,** The thermal regime of smoke formation in smoke generators, *Rybn. Khoz.,* No. 3, 80, 1974.
38. **Golovanev, V. F. and Bolshakov, A. S.,** Effect of cooking and smoking methods on the quality of textured products, *Izv. Vyssh. Uchebn. Zaved. Pishch. Tekhnol.,* No. 3, 71, 1978.
39. **Alexander, A.,** Electrostatic principles applied to smoking technology, *Food Flavor Ingred. Pkg. Proces.,* 1(5), 31, 1980.
40. **Baltes, W., Wittkowski, R., Sochtig, I., Block, H., and Toth, L.,** Ingredients of smoke and smoke flavor, in *The Quality of Foods and Beverages,* Vol. 2, Charalambous G. and Inglett, G., Eds., Academic Press, New York, 1981, chap. 1.
41. **Hollenbeck, C. M.,** Novel concepts in technology and design of machinery for production and application of smoke in the food industry, *Pure Appl. Chem.,* 49, 1687, 1977.
42. **Wistreich, H. E.,** Smoking of meats, Proc. Meat Ind. Res. Conf., Chicago, 1977, 37.
43. **Baichev, I. and Stefanov, Y.,** Application of liquid smoke for the production of cooked ham, *Proc. Eur. Meet. Meat Res. Workers.,* No. 25, 12.13:945, 1979.
44. **Wesley, P.,** Liquid smoke stays brine-soluble, *Food Prod. Dev.,* 16(2), 30, 1982.
45. **Keller, G. and Gunther, H. O.,** Investigations into aroma compounds of seasonings treated with liquid smoke, *Fleischwirtschaft,* 56(4), 564, 1976.
46. **Potthast, K.,** Using liquid smoke for surface treatment of meat products, *Fleischwirtschaft,* 63(10), 1558, 1983.
47. **Schneck, J. C.,** Liquid smoke application to cured meat, *Proc. Annu. Rec. Meat Conf. Am. Meat Sci. Assoc.,* 101, 1982.

48. **Hsu, L. A.**, Technology of smoking meat products, *Boletin Inst. Technol. Aliment. Brazil*, No. 52, 1, 1977.
49. **Gubler, D. and Vernois, M.**, New smoking techniques, *Ind. Aliment. Agric.*, 89, 621, 1972.
50. **Wechel, L.**, Optimum smoking methods, *Fleischwirtschaft*, 62(10), 1264, 1982.
51. **Hsu, L. A. and Sink, J. D.**, Color of meat products with liquid smoke products, *Coletanea Inst. Technol. Aliment.*, 8, 271, 1977.
52. **Rusanov, V. V.**, Spectrophotometric determination of phenols in smoking liquids, *Izv. Vyssh. Uchebn. Zaved. Pishch. Tekhnol.*, No. 1, 50, 1976.
53. **Baltes, W., Wilcopolski, K., and Lehmann, B.**, Detection of liquid smoke preparations in sausages, *Lebensm. Gerichtl. Chemie*, 33(4), 73, 1979.
54. **Baltes, W. and Bange, J.**, The detection of liquid smoke aromas smoke flavorings in foodstuffs, *Acta Aliment. Pol.*, 3, 325, 1977.
55. **Pionov, I. T., Marinosyan, O., Kuyumdzhiev, K. N., Nenchev. N. R., and Khodikyan, N.**, Smoke generator with automatic regulation of smoke formation, *Proc. Eur. Meet. Meat Res. Workers*, 31, 725, 1985.
56. **Brazhnikov, A. M., Kamenskii, S. N., Malova, N. D., Babanov, I. G., Sukhanova, S. I., and Minaev, A. I.**, Distribution of air in thermal chambers for processing cold smoked sausages, *Myasn. Ind. SSSR*, 4, 39, 1985.
57. **Lange, W. G. K.**, Calculation method of smoking technology, *Fleischerei*, 37(2), 88, 1986.
58. **Simmonds, C. K., Avery, K. W. J., and Higgo, D.**, Smoking of rainbow trout, *Annu. Rep. Fish. Ind. Res. Inst. Cape Town*, 38, 8, 1984.
59. **Simmonds, C. K. and Avery, K. W. J.**, Smoked and canned catfish and Kurper, *Annu. Rep. Fish Ind. Res. Inst. Cape Town*, 38, 8, 1984.
60. **Maharova, N. A., Goncharova, A. M., Pasternak, V. E., and Kosinova, R. N.**, Pilot plant experiments with smokeless cold smoking of fish, *Rybn. Khoz.*, 4, 56, 1985.
61. **Kurko, V. I. and Mezenova, O. Y.**, Determination of completeness of precipitation of smoking preparation during electrostatic smoking of fish, *Rybn. Khoz.*, 9, 58, 1985.
62. **Kurko, V. I. and Mezenova, O. Y.**, Diffusion of phenolic compounds into fish during smokeless electric smoking, *Rybn. Khoz.*, 7, 70, 1985.
63. **Makarova, N. A., Goncharova, A. M., and Tatarenko, L. Y.**, Microbiological study of cold smoked fish during storage, *Rybn. Khoz.*, 1, 70, 1985.
64. **Toth, L. and Potthast, K.**, Chemical aspects of the smoking of meat and meat products, *Adv. Food Res.*, 29, 87, 1984.

Chapter 4

FLAVOR CHEMISTRY OF WOOD SMOKE

I. INTRODUCTION

If a general population who is familiar with smoked foods were asked what they liked best about smoked foods, the majority would undoubtedly answer that is was the unique flavor. Thus, it is imperative that the flavor composition of wood smoke be understood so that uniform optimum flavor can result.

Numerous researchers have devoted a great deal of time and effort attempting to actually identify, and in a few cases quantitate, the compounds that are present in wood smoke. Considering the relatively few number of constituents present in most woods, this would seem to be a rather simple and straightforward task, but as anyone who has been exposed to flavor chemistry knows, the organoleptic properties of most products are far from simple.

Wood smoke is no exception, and this is compounded by the fact that numerous other factors typified by wood moisture content and temperature of smoke generation can also dramatically influence wood-smoke chemical composition. In addition, certain compounds apparently only serve as intermediates in the formation of more stable forms of compounds. Also, in most model systems involving smoke generation not in the presence of food, one is usually only concerned with volatile compounds that may or may not have distinctive odor properties at low concentrations. This in turn does not account for the fact that certain nonvolatile compounds may also make significant contributions to flavor. These facts are further complicated when food is actually exposed to volatiles and particulate matter found in smoke because these compounds in turn can chemically react with certain food components to form an entirely new generation of flavor compounds.

Fortunately for the flavor chemist, the identification of compounds present in wood smoke has been made somewhat easier through the use of combined gas chromatography/mass spectrometry. Most of the over 400 volatiles identified to date in smoke and its by-products have used this technique. For your information, most of the compounds identified to date have been summarized in Table 1.* This list includes 48 acids, 22 alcohols, 131 carbonyls, 22 esters, 46 furans, 16 lactones, 75 phenols, and 50 miscellaneous compounds. As always, when compiling an extensive list of this type, oversights may exist. In any event, it should be fairly apparent that what most people consider to be something relatively simple, in actuality can be quite complex.

It should also be noted that just because a compound has been identified as being present in smoke does not necessarily mean that it makes any contribution to the overall or certain specific sensory properties of the resulting product. Aside from its presence, one needs to know the following before the contribution of a compound to flavor can be fully evaluated: the quantity present; its odor threshold or minimum concentration required to be perceived; the specific odor properties or descriptions of the compound in question over a wide concentration range; potential synergistic odor properties of the compound with similar compounds in a homologous series, stability to temperature, light, and oxygen; and possible chemical interaction with other compounds. If one were to consider all of these factors in evaluation of the 400-plus compounds listed in Table 1, a great deal of time and effort would result. Therefore, it is not surprising that to date the individual compounds that are totally responsible for the characteristic flavor and aroma of wood smoke have not been completely and positively identified.

* Tables appear following text.

This becomes even more frustrating when one realizes that the compounds responsible for the characteristic flavor associated with different wood sources has not even been resolved. Therefore, in general, flavor chemists currently conclude that the overall characteristic flavor of wood smoke is probably due to the contribution of compounds from several compound classes, each of which contribute certain portions to the overall sensory properties of wood smoke.

At this point, perhaps we should discuss in some detail the classes of compounds that are thought to contribute most to the flavor and odor properties of wood smoke.

II. IMPORTANT FLAVOR COMPOUND CLASSES

A. Phenols

Historically numerous researchers[2,3,5,7,9-11,17,19,25,29,31,40,42] have concluded that phenols, as a compound class, are the primary contributors to wood-smoke aroma. However, other groups dispute this claim[1,4,8,11-16,18,20] since they have found other classes of compounds that also produce typical smoke aroma properties. Probably the primary reason for this conflict is the fact that for the most part, rather primitive and thus usually incomplete fractionation procedures have been utilized to recover what is normally called the phenolic fraction, which actually contains other compound classes which also can have potent sensory properties. This classical isolation procedure is shown in Table 2. If the fractions named in Table 2 are subjected to gas chromatography, one can readily see the problem, since several types of compound classes are present in each fraction, and if the sensory properties of each named fraction are evaluated, erroneous conclusions can be made. What is actually needed in this area are more sophisticated forms of separation to insure the purity of each fraction.

Structurally, phenols are aromatic hydrocarbons composed of benzene to which varying numbers of hydroxy groups are attached. In addition, phenolic compounds can have other functional groups such as aldehydes, ketones, acids, and esters. Simple phenols, when dissolved in water, display acidic properties while in alkaline conditions they are solubilized by dissociation to form phenolates. Increasing the number of hydroxy groups present on the benzene ring causes the resulting phenol to be quite reactive with metals and sensitive to light and oxygen. A monohydroxyphenol has a boiling point of 183°C whereas the addition of a second hydroxy group increases the boiling point to 270°C.[73] Ether forms of phenols are insoluble on water and can be converted to phenol ketones at low temperature or can form condensation products by reacting with aldehydes.

Not considering wood type or combustion conditions, there is no question that phenols can be produced at relatively high levels in wood smoke. The major ones found by Kornreich and Issenberg[3] are shown in Table 3, and as can be seen, the concentration of phenols in whole smoke can be 10 to 60 times that in the vapor phase of some smoke products. Thus, type of smoke application utilized would in turn influence phenolic concentration, and thus in turn, product flavor. For example, Ishiwater et al.,[12] in surveying 15 different samples of commercial smoke flavors, found a total phenolic range of from less than 0.001 to 1.06 mg/mℓ.

Lustre and Issenberg[5] demonstrated that not all phenolic compounds produced in wood smoke end up as such in smoked foods. These include the compounds 2,6-dimethoxy-4-vinylphenol, 2,6-dimethoxy-4-*cis* and *trans* propenylphenol, propiosyringone, coniferaldyhyde, and sinapaldehyde. This is interesting in light of the fact that especially the last two compounds are present in wood smoke at fairly significant levels. Because their structures contain carbonyl groups and an activated double bond, they are prone to addition reactions, and thus, the authors concluded that they can react with meat proteins to form other compounds. However, another research group[11] did find sinapaldehyde in a variety of smoke foods, thereby indicating that apparently some of the compound can be associated with certain smoked foods in the unreacted state.

It should also be noted that with storage, the amount of individual free phenols can decrease dramatically. Data supporting this are summarized in Table 4. It was clearly shown that volatility was not a contributing factor. Probable covalent binding of quinones, resulting from the oxidation of phenols to reactive protein groups probably resulted.

Another factor to consider relative to phenols in foods is their degree of penetration into smoked foods. As would be expected, there is preferential adsorption on outer surfaces. Bratzler et al.,[50] in working with a smoked bologna emulsion, clearly demonstrated this. A portion of their data is summarized in Table 5.

Consideration should also be given to the fact that the fat content of a smoked food can influence the amount of individual phenol compound recovered. For example, Issenberg et al.,[51] in working with model systems, demonstrated, as can be seen in Table 6, that for some phenols, percent recovery was lower in water than from a lipid system, whereas for some compounds it was essentially the same (guaiacol), while for others, higher recovery was recorded for the lipid system. Thus, depending upon the levels of individual phenols and the amount of fat in the food, the amount of total measured phenols can vary.

In the unique case of utilizing fire-charred wood barrels for the storage and aging of distilled spirits, one should also consider the fact that compounds formed during the partial thermal destruction of wood can then be dissolved and adsorbed into the final product. In the case of phenolic compounds, Maga[68] has identified 28 such compounds. These are summarized in Table 7 and as can be seen, all of the phenols are common compounds relative to wood smoke. In addition, a series of 13 phenolic acids, some of which are chemically quite active, was also identified.

1. Sensory Properties

Numerous researchers have attempted to evaluate the role of individual, as well as the composite phenol fraction, to the sensory properties of wood smoke. As can be seen in Table 8, certain phenols can have fairly low taste and odor thresholds, but as pointed out by Wasserman,[52] one should also consider the actual amount of compound present. Therefore, if you look at what Wasserman[52] calculated to be the flavor index, by far the most important phenol was 4-methyl-guaiacol. It can also be seen that certain phenols had lower odor thresholds in a water system, whereas, others were more sensitive in an oil system, thus making it difficult to make generalizations as to the overall odiferous properties of phenols in a smoked food.

Wasserman[52] also asked a taste panel of approximately 20 people to describe the odor and taste properties of the three phenols shown in Table 9, and as can be seen, the major response for all three compounds was smoky, with the second major response being phenolic. The third major response varied dependent upon the sensory properties being evaluated. In the case of odor it was sweet, while for taste it was bitter. The compound found to have the most smoky odor sensation was 2,6-dimethoxyphenol.

Since phenols are significant contributors to smoke aroma, data have been reported[73] indicating that phenols of medium volatility are the most important. For example, the flavor properties of phenol fractions distilled from a liquid smoke were characterized as follows: the low-boiling fraction (60 to 90°C) composed primarily of phenol, cresols, guaiacol, and methyl-, and ethylguaiacol had a hot and bitter taste; the medium fraction that distilled over at 91 to 132°C and contained *cis*- and *trans*-isoeugenol, syringol, and methylsyringol had a pure and characteristic smoke flavor; the high-boiling phenol fraction (133 to 200°C) had an acid, chemical property that was judged of poor quality.

This in turn probably points out a void in the published data relative to the sensory properties of phenols, in that their odor has been described in somewhat general terms, and usually these compounds are only evaluated in relatively concentrated forms. This is exemplified by a portion of the data reported by Kim et al.[4] as summarized in Table 10. Here you can see general sensory terms such as pungent, cresolic, and burnt. Similar data have been reported by Baltes and Sochtig.[43] Contrast the rather general sensory descriptions for phenols in Table 10 with the very detailed descriptions for some of the same compounds in Table 11. You can also note in Table 11 the approximately tenfold differences in optimum sensory concentrations for these compounds. A most useful study would be to describe the odor properties of increasing concentrations of each individual compound and then to evaluate the potential synergistic effect by combining some of the more interesting compounds.

Some data based upon the lines of the above suggestions have been reported.[73] The sensory properties of increasing concentrations of individual phenols were characterized by a panel and as seen in Table 12, the concentrations of each phenol that produced an optimum response was varied with dimethylphenol being needed in the lowest concentration and isoeugenol in the highest. Also, all compounds showed smoke-like properties along with other sensations. Data of these type can serve as the basis for the formation of imitation smokes.

The fact that phenols are important to overall whole smoke aroma should not be considered to be the only class of compounds involved was nicely demonstrated by Fujimaki et al.[29] Some of their data are shown in Table 13. They fractionated whole smoke condensate and then evaluated the sensory effect of adding these fractions back to the phenolic fraction, and as can be seen, the combination of at least three fractions produced the best results. However, it should be noted that the two different combinations of three fractions produced the best results although their aroma descriptions were different. Thus it would appear that depending on the overall characteristic, and perhaps intensity, of smoke aroma desired, different fractions are involved.

Olsen[42] also fractionated a commercial liquid smoke preparation and with the aid of a sensory panel concluded that the phenol fraction was important but not complete from a sensory standpoint. These data are summarized in Table 14. It can also be seen that normally objectionable tastes such as chemical and acidulous can be associated with certain fractions but were not apparent in the composite sample.

In an interesting study, Daun[53] isolated the phenolic fraction from both the vapor and particulate phases of wood smoke and through dilution determined, with the aid of a sensory panel, their recognition threshold and most desirable concentration for both odor and taste sensations. These data are summarized in Table 15. For both odor and taste the recognition threshold was lower in the particulate phase as compared to the vapor phase, thus indicating that either a higher level or more potent phenols were associated with the particulate phase. Because of this, the most desirable concentration for both odor and taste was also lower in the particulate phase, in turn indicating that if smoke containing a particulate phase is being used, a smaller amount of smoke is required to achieve an optimum level of smoke odor and taste.

2. *Formation Pathways*

It has been clearly demonstrated by Fiddler et al.[41] that a large number of phenols can result from the key thermal degradation product of lignin, ferulic acid, which in turn can form another key intermediate, 4-vinylguaiacol. The conversion of ferulic acid to 4-vinylguaiacol is presented in Table 16. Maximum numbers of by-products result in the presence of air, although some by-products will also form in the presence of nitrogen. Some of the typical by-products isolated from the thermal degradation of ferulic acid are shown in Table 17 and those that are thought to form from 4-vinylguaiacol are shown in Table 18. Some of these by-products in turn can undergo further chemical reaction to form other compounds.

In an earlier chapter it was shown that lignin is composed of repeating phenol units having three-carbon sidechains. Also, lignin from softwoods has fewer methoxy groups than lignin from hardwoods which accounts for the different range of phenols found in smoke generated from these different types of woods. However, lignin is not the only source of phenols since they can also be derived from cellulose via the conversion at high temperature of aliphatic compounds to aromatic compounds.[73] Also, phenols can result from the thermal degradation of amylopectin which is a component of hemicellulose.

One should also not forget that spices and their thermal decomposition can be a source of phenols in smoked foods. Some of the common spices used with smoked foods and the phenols associated with each are summarized in Table 19, and as can be seen, a wide range of phenols are associated with these spices.

B. Furans

Furans represent a class of five-membered oxygen-containing heterocyclic compounds that most probably result from the dehydration of glucose which in turn is an intermediate of the thermal dehydration of cellulose. Other furans such as furfuraldehyde can be formed from pentosans which are degradation products of hemicellulose. Furfuraldehyde in turn can condense to form numerous analogs that possess aromas such as sweet, fruity, and grassy.[73] The major six-membered ring oxygen-containing heterocyclic compound found in smoked foods is maltol.

Kim et al.[4] feel that furans also contribute to the overall sensory properties of wood smoke in that they tend to soften the heavy smoky aromas usually associated with phenolic compounds. Some of the odor properties associated with a portion of the furans isolated from wood smoke are summarized in Table 20. Radecki et al.[17] have also reported that a furan fraction isolated by the steam distillation of a commercial liquid smoke, in which six furans were identified, possesses a sweet, fragrant, and floral aroma, which also was felt to soften heavy smoky aromas. Numerous furans were also identified by Maga[68] in charred oak and are summarized in Table 21.

C. Carbonyls

Carbonyls are also thought to contribute to the overall sensory properties of wood smoke. For example, Fiddler et al.[1] isolated a polyfunctional carbonyl subfraction from wood smoke that possesses a caramel or burnt sugar aroma which the author felt modified the flavor of phenols.

Some of the carbonyls isolated by Kim et al.[4] from wood smoke were found to have unique aroma properties, as can be seen by viewing Table 22. Most had an odor described as grassy, which the authors contributed little to overall aroma, but no data were provided to substantiate this claim.

Most carbonyls are thought to originate through the thermal decomposition and rearrangement of cellulose and hemicellulose via classical carbohydrate degradation schemes.[4] However, in going back to Table 1 of this chapter, it should be noted that the carbonyl class represented the largest number of compounds identified to date from wood smoke. Since this general class represents a diverse group of structural compounds, more effort should be made to more completely understand their importance to wood-smoke aroma.

A carbonyl of potential concern is the compound formaldehyde, due to its toxicity. It has been found at levels of up to 200 mg/100 g of wood in smoke and at a concentration of up to 50 mg/kg in smoked meats.[73]

Certain ketoalcohols can also act as aliphatic intermediates for the formation of heterocyclic hydroxyfurans and alicyclic compounds such as cyclopentanones which possess unique and potent burnt maple aromas.

D. Others

Two other groups of flavor compounds that have received relatively little attention as to their potential role in wood-smoke flavor chemistry are the lactones and pyrazines. As can be seen in Table 23, a rather extensive group of lactones has been identified by Kim et al.[4] and all possessed aroma properties that can be considered to be characteristic[71] and beneficial to wood smoke aroma. The lactone fraction had a yield of 29.2 mg% with approximately half (14.5 mg%) being due to the compound 2,4-dimethyl-2-butenalide, with the compound 2-ethyl-4-methyl-2-butenalide contributing 8.1 mg%. They postulated that lactones in wood smoke are primarily derived from the thermal degradation of cellulose and hemicellulose via the intermediates butyrolactone and 2-butenolide. These compounds in turn form hydroxyacids. For example, the dehydration of α-hydroxybutric acid results in the formation of γ-butyrolactone, while the dehydration of α-crotonic acid produces 2-butenolide. These end-products in turn can serve as intramolecular esters needed for the conversion of aliphatic compounds to aromatic ones.

Several reports have appeared identifying pyrazines in wood smoke or charred wood. Pyrazines are characterized as heterocyclic nitrogen-containing compounds that are primarily thermally produced and possess potent and unique sensory properties.[65-67] Kim et al.[4] found the compounds pyrazine, methylpyrazine, and 2,5- (or 2,6-) dimethylpyrazine at a combined level of 1 mg% in the basic fraction of wood smoke. Since wood contains a nitrogen source, it is possible that a large number of these compounds are actually present in wood smoke.

Recently, Maga[68] reported on the volatile composition of fire-charred white oak (*Quercus alba*). Charred oak was fractionated into basic, carbonyl, phenolic, and neutral portions and their sensory contribution and chemical composition determined. Based on a 10-point intensity scale, with 1 being bland and 10 intense, a sensory panel concluded that the basic fraction, whose composition is shown in Table 21, had an intensity rating of 7.8, which was followed by the carbonyl, neutral, and phenolic fractions at 7.2, 6.8, and 6.2, respectively. Thus, based on both seonsory and objective data centering on pyrazines, he concluded that in the case of charred white oak, pyrazines are important sensory contributors and probably find their way into alcoholic beverages that are subsequently stored in charred oak casks. Pyrazine composition in wood smoke as influenced by wood source will be discussed in a later section.

In addition, numerous types of up to five-carbon aliphatics such as hydrocarbons, alcohols, and acids are present in smoke and some smoked foods. Of special interest is the toxicologically active compound methanol which is primarily derived from the methoxy groups of lignin. However, upon oxidation it can either form formaldehyde, which also presents toxicological concerns, or formic acid.[73]

III. WOOD SOURCE

It is a well-accepted fact that the type of wood used for a smoke source can significantly influence the organoleptic properties of the resulting smoke. Because of availability, cost, and numerous other factors, many types of wood have historically been used. In general, it can be stated that wood from broad-leaved trees produces a more acceptable smoke than smoke derived from needle-leaved trees. However, this may not always be the case. As can be seen in Table 24, pine was judged to produced a better smoke than fir, while some broad-leaved woods produced as inferior a smoke as fir.

Jodai[28] evaluated the effect of pyrolignous acid extraction from charcoal derived from cherry, oak, and pine using a variety of solvents including benzene, butyl acetate, hexane, petroleum ether, ethyl ether, and trichloroethylene. Benzene and butyl acetate were found to be the most efficient solvents with cherry and oak-derived charcoal giving acceptable-smelling products. The compounds that were thought to contribute most to a desirable product were cresol, 2,6-dimethoxyphenol, methyl ethyl ketone, and 3-methyl-2-butanone. However, pine-aroma extract was found to be unacceptable with all solvents.

Probably the most extensive study centering on the influence of wood source on resulting aroma acceptability and composition was conducted by Fujimaki et al.[29] Wood sources that were utilized included two oak species (*Queretis acuta* and *Q. seriata*), cherry, bamboo, pine, and cedar. Some condensate generated from each wood source was distilled and further fractionated into basic, carbonyl, noncarbonyl, and phenolic acid portions. The sensory and chemical composition of each fraction was then evaluated. As can be seen in Table 25, the total and individual amounts of volatile materials varied significantly among wood sources. For example, cedar had approximately half the amount of total volatiles as the other wood sources, most of which were composed of acids. The amount of phenolic material was fairly constant in all wood sources, while the carbonyl, noncarbonyl, acid, and basic fractions varied dependent upon wood type. Interestingly, *Q. acuta* and cherry had proportionately high yields of the basic fraction, which exhibited a strong burnt aroma. The overall flavors of the smoke condensates were reported to be distinctly different with *Q. acuta* having a strong burnt flavor with some sweetness; *Q. seriata* having a heavy, sweet smoke which was judged to be best of all woods evaluated; bamboo also had a pleasant smoke flavor; pine and cedar had pungent phenolic and poor fragrance sensory properties; while cherry was variable with some samples possessing oak-like properties while others were more pine-like.

Through organoleptic evaluation, the authors reported that the phenolic fraction had a good characteristic smoky aroma that was monotonous while the phenolic/carbonyl fractions had a better fragrant smoke aroma. The combination of the phenolic and neutral fractions had a very good aroma that was described as possessing a complex and fragrant property. The most pleasant aroma resulted from the combination of phenolic, neutral, and basic fractions. This product was thought to closely approximate the original smoke aroma, but without an acid odor. Thus, the authors concluded that although the phenolic fraction of wood smoke is an important contributor to smoke aroma, other fractions tend to complement and add complexity to smoke aroma.

Gas chromatographic analysis of the major volatile compounds associated with the whole flavor concentrate showed no major differences, however, when each fraction was chromatographically evaluated, differences were apparent among wood sources in that certain minor constituents were either present or absent. Thus, even though the same major compounds were apparently present in all woods, flavor differences were apparently due to differences in the relative amounts of these compounds. A specific example of this was the fact that syringol and some of its substituted forms were present in broad-leaved wood sources but absent in needle-leaved woods. Also, the ratio of guaiacol and its derivatives, as compared to phenol, cresols, and xylenols, appeared to be different in the smoke of broad- and narrow-leaved tree woods. Actual compositional differences of the compounds identified from the various wood sources are shown in Table 26.

Baltes et al.[64] also presented some interesting similar data, which are summarized in Table 27 showing that relative to phenols the major difference between hardwood and softwood is the higher proportion of 4-methylguaiacol and the lower proportion of syringol, 4-methylsyringol, and 4-propenylsyringol in softwood as compared to hardwood. Based on this fact, Baltes et al.[64] proposed that it is possible to determine if hardwood or softwood was used as a wood smoke source simply by determining the proportion of syringol to guaiacol and its derivatives.

In comparing the carbonyl composition of smoke derived from either alder or fir, Borys[30] found that fir contained more aliphatic ketones, cyclic ketones, and aromatic carbonyl compounds than alder. However, the potential sensory significance of this was not discussed.

In the case of fir-wood smoke, a total of 39 compounds were identified while 28 compounds were found in alder wood smoke. Khvan et al.[56] also found major volatile compound differences in smoke as dictated by wood source. Smoke from oak was found to contain more volatile acids, neutral, and basic compounds compared to pine-derived smoke. However, due to the larger amount of resin present in pine, more phenols were found in pine smoke.

Lee and Song[57] compared various chemical and sensory properties of oak and chestnut smokes. One factor that would tend to indicate differences between the two woods was pH. Oak smoke had a pH of 2.20 while the pH of the chestnut smoke was 2.65. However, when these two smokes were incorporated into a sausage, a taste panel showed no statistical difference between the two. The actual volatile composition of the two smokes was not evaluated.

Ziemba[72] reported that smoke from oak wood contained more organic acids than beech wood and thus was inferior to beech. His data comparing the smoke composition of these woods when used for light and heavy smoking are summarized in Table 28.

In general, it has been reported[39] that woods such as beech and oak, which are considered to be hardwoods, produce smoke that is high in acids, while softwoods, such as pine and spruce, have high levels of carbonyls in their smoke. Also, poplar is quite high in phenols. Data supporting this observation are summarized in Table 29.

Several research groups have evaluated the influence of wood type on the resulting sensory properties of smoked fish products. In one study, softwoods such as fir and red cedar produced a resinous flavor in smoked herring,[58] however, other workers[55] found that sprats smoked with pine were better than those using beech wood.

An extensive study with smoked white fish, using 12 wood sources, was conducted by Lantz and Vaisey.[59] Using a 61-member taste panel, the data in Table 30 were tabulated. They found that red maple, red oak, trembling aspen, white ash, balsam, poplar, and birch gave better-flavored fish. Beech and white oak only gave moderately acceptable products, while hard maple and hickory were only marginally acceptable and diamond willow and burr oak resulted in objectionable products. Actual chemical analyses on volatile compound differences among the woods were not performed. It is of interest to note that hickory wood was not rated very highly for fish since in the U.S. it is thought to be one of the most acceptable woods for the smoke processing of red meats. The sensory and/or chemical reasons for this difference should be investigated. Perhaps because of differences in protein content or composition between fish and red meats, different types or amounts of flavor compounds are formed.

Solanki et al.[60] smoked eel, also using various wood sources and noted the resulting color, odor, and flavor. As can be seen in Table 31, overall, the best wood source was a blend of one part coconut husk and one part teak sawdust. Using teak sawdust also produced an attractive color, but the odor and flavor were acrid and unpleasant. The use of sajad, acacia, and mango sawdust produced a resinous odor and flavor. In contrast, coconut husk has a good odor and flavor, but the resulting color was a pale yellow. Therefore, by blending wood sources they were able to optimize sensory properties. This concept of blending woods has not been investigated extensively, but it would appear that with certain smoked foods, a very complex combination of sensory properties could be achieved, thereby adding another dimension to the acceptability of smoked foods.

As mentioned earlier, numerous wood and related by-products have commercially been used as a smoke source, and there is no question that their resulting flavors can vary widely. Some of these sources include mahogany, numerous fruit tree woods, corn cobs, prunings from grape vines, and more recently, mesquite wood.

Recently, Maga and Fapojuwo[69] and Maga and Chen[70] reported upon the sensory and volatile pyrazine composition of various wood smoke sources. Smoke generated from apple, chestnut, cherry, hard maple, hickory, mesquite, and red and white oaks were fractionated into carbonyl, neutral, and basic and phenolic units. Equal concentrations of each were dissolved in water and presented to a sensory panel for aroma intensity evaluation using a 10-point scale with 1 being slight and 10 strong. As seen in Table 32, it is clearly demonstrated that the fraction possessing the strongest aroma intensity is dependent upon wood source. For example, in the case of mesquite, the phenolic fraction appears to be most important whereas with apple, the basic fraction has the strongest intensity. These data would support the concept of blending wood smoke sources to obtain complex aroma properties as discussed above.

A series of eight pyrazine compounds including pyrazine, 2-methoxy-3-ethylpyrazine, 2-ethoxy-3-ethylpyrazine, 2-propyl-3,6-dimethylpyrazine, 2-butyl-3,5-dimethylpyrazine, 2-butyl-3,6-dimethylpyrazine, 2-acetyl-3-methylpyrazine, and 2-acetyl-3,5-dimethylpyrazine were identified in the smoke generated from 17 different wood sources[70] and as seen in Table 33, the total pyrazine content ranged from a low of 10.33 mg/100 g of redwood to 41.03 mg/100 g of hickory. There did not appear to be any obvious differences between hardwoods and softwoods. Also, total and individual pyrazine levels were not found to be correlated with wood nitrogen content, which has to be assumed to be the nitrogen source, nor to the levels of cellulose, hemicellulose, or lignin present. Again, the authors concluded that based on both the amounts and types of pyrazines found, they can contribute to overall smoke aroma.

IV. GENERATION CONDITIONS

One would assume that if variables in the combustion process such as temperature, moisture, and air levels are manipulated, the amount of flavor-producing compounds can vary, thus in turn influencing the overall sensory properties of the resulting smoke. Numerous studies have demonstrated this, but just a few typical ones will be cited.

A. Temperature

Relative to temperature, data presented by Porter et al.[18] as summarized in Table 34 are fairly typical of what happens to the quantities of flavor compound classes. As can be seen, there is an increase as temperature increases, and then a general decline at higher temperatures.

Daun[53] generated smoke at higher temperatures than the above study and although the actual 14 phenolic compounds were not identified even though they were reported gas chromatographically, general increases and decreases were evident for these compounds as temperature was increased from 400 to 600°C. In turn, this had a significant influence on the vapor phase recognition threshold and most desirable concentration for both their odor and taste properties. This information is summarized in Table 35. As can be seen, the smoke product at 400°C apparently was more concentrated in flavor compounds than at the other two temperatures since both its odor and taste thresholds and most desirable concentrations were lower.

The amount of phenolic material generated at still higher temperatures has been reported by Toth,[27] and as can be seen in Table 36, the same general trend of an increase with gradual decline at the highest temperature is evident. Baltes et al.[64] also reported that maximum phenol, carbonyl, and acid yields occurred at approximately 600°C. They also stated that double-valanced phenols, such as pyrocatechin, are especially stable at high temperatures, therefore increasing their relative amount at the expense of thermally less-stable compounds.

Simon et al.[61] looked at compound compositional change associated with small smoke temperature generation increases and they found, as can be seen in Table 37, the actual weight percent of acids and phenols decreased with increasing temperature, while carbonyl content increased. However, the total number of volatiles increased by a factor of 4.5 over the 115°C temperature range evaluated. Since the relative amounts of acids and phenols were decreasing with increasing temperature, it becomes apparent that the relative amount of carbonyls increased significantly.

If the actual content of individual phenols is looked at relative to smoke generation temperature, some interesting facts appear. For example, in Table 38 it can be seen that for the compound phenol, temperature makes little difference in the amount formed, but for syringol an increase of 200°C from 450 to 650°C resulted in almost a fourfold increase. Thus, it would appear that at least in a theoretical sense one can encourage the generation of certain phenolics by adjusting generation temperature.

Borys et al.[24] attempted to mathematically and chemically evaluate various combustion conditions, including temperature, on the resulting volatiles. Their initial assumption was that combustion rate was dependent upon combustion temperature and air flow rate relative to the equation:

$$Y = A + B\,(T \times u)$$

where Y = combustion rate, A = constant dependent on wood particle size, B = constant dependent on wood moisture content, T = combustion temperature, and u = air flow rate.

Based on their experimental smoking conditions, very high correlation coefficients were found for this formula. Later work reported by Klossowska[62] also used the same mathematical approach and she concluded that the logarithm of the wood combustion rate is a linear function of the combustion temperature and the logarithm of the air flow rate. In addition, she concluded that the temperature of the combustion zone is limited to the rate of air flow by a functional relationship representing a Gaussian curve.

In evaluating variables that influence the types and amounts of pyrazines formed during wood smoke generation, Maga and Chen[70] reported that the concentration of eight pyrazines increased from 41.03 mg/100 g dry weight wood when hickory sawdust was thermally degraded at 290°C to a total of 59.05 mg/100 g when a temperature of 450°C was used.

One should remember however that as pointed out by Wasserman and Fiddler,[39] the actual measurement of temperature at the combustion site is somewhat difficult. For example, they found that the temperature 1 in. above a smoldering sawdust bed was as much as 300°C lower than in the center of the bed. In addition, wood combustion is an exothermic reaction, and therefore, localized sites may have temperatures that are far greater than the composite temperature or temperature of the heat source.

B. Air

The amount of air present during smoke generation is another factor that can apparently influence the resulting flavor component composition. As an example, Wasserman and Fiddler[39] varied oxygen composition from 0 to 59% of the gas mixture used during smoke generation and saw some major differences in the volatile composition of several measured compounds as a result. These data are shown in Table 39. Normal air has an oxygen content of 20%. In general, they found that as percent oxygen increased, furans, typified by the compound furfuryl alcohol, decreased, thus demonstrating that oxygen is not prerequisite for the formation of furans. Interestingly, phenolic compounds (guaiacol) and carbonyls (cyclotene) increased significantly at 10% oxygen, but then also fell off as oxygen content increased.

With the aid of a sensory panel, Wasserman and Fiddler[39] concluded that the sensory smoke properties of condensates derived from 0 to 10% oxygen could be statistically differentiated from 20% oxygen, but no statistical differences were noted when 20, 30, and 50% oxygen-produced condensates were compared. Thus, at least in this study, smoke condensate produced with normal amounts of oxygen was found to be quite acceptable. The question is raised, however, if this result is due only to the fact that this produced the most normal sensation and that lower levels of oxygen produced different and perhaps therefore not as agreeable condensates.

A sensory panel was also used by Daun[53] to compare the sensory properties of the vapor phase of smoke produced in the presence of low (500 ℓ/hr) and high (2500 ℓ/hr) air flow. As seen in Table 40, smaller amounts of the smoke produced with low air supply were needed for odor recognition threshold and most desirable odor concentration as compared to the high air level. This would be in general agreement with the Wasserman and Fiddler[39] study which found higher amounts of certain flavor-producing compounds at lower oxygen levels. However, when taste was considered, the smoke produced with high air levels required less to be most desirable, perhaps indicating that a wider variety of compounds or compound concentration was a factor that should be considered.

The presence or absence of air significantly influenced the amount of pyrazines formed in hickory smoke.[70] With air, the total amount was 41.03 mg/100 g of wood, however, when air was excluded during smoldering, the total pyrazine content increased to 51.33 mg/100 g. In all probability, the lack of air minimized formation of certain classes of unstable compounds thus resulting in larger amounts of more stable pyrazines.

C. Wood Moisture Content

One would also assume that initial wood moisture content could possibly alter the type and amount of volatile compounds in smoke, and this in turn would influence overall sensory properties. Data published by Gorbatov et al.[63] clearly support this concept. As shown in Table 41, lower moisture wood tended to produce higher amounts of phenols, acids, and carbonyls, but these data should be interpreted carefully since it can be seen that the total amount of condensate also varied. If the data were calculated on the same amount of condensate yield, the middle moisture samples would have proportionately more volatiles. Interestingly, the middle moisture samples were those that were judged to have the most acceptable sensory properties. Since the formation of certain flavor compounds is moisture dependent, more research should be devoted to determine which specific compounds are preferentially formed in low and high moisture-containing foods during smoking.

Recently, Maga and Chen[70] followed the rate of pyrazine formation as influenced by hickory wood moisture content at the time of smoldering at 290°C. Through moisture additions, they generated hickory smoke from sawdust that had either 4, 20, or 30% moisture. No significant differences in total pyrazine contents were noted when the sawdust had 20 or 30% moisture, but pyrazine concentration doubled when the moisture content was at 4%. Thus, if one were interested in maximizing the generation of pyrazines in wood smoke, one should use a fairly dry wood source for smoke generation.

D. Wood Particle Size

The actual size of the wood particle being used as a smoke source can indirectly influence resulting smoke flavor in that, as pointed out by Borys et al.,[24] the larger the particle, the higher the air flow rate required to obtain the same combustion rate as a smaller particle. Thus, if air flow is not altered according to particle size, combustion rates will differ, which in turn will influence final smoke composition. Again, this is an area that deserves further research in that most naturally occurring sawdust a wide range in particle size can exist, thereby possibly leading to variations in volatiles produced.

V. SMOKING PROCEDURE

As discussed in an earlier chapter, numerous smoking procedures exist, some of which claim that superior sensory properties result. This in turn would indicate that some unique aspect of the process, probably including some or all of the factors discussed above, results in the formation of more or a wider range of flavor compounds.

Data typical of this contrast are presented in Table 42 where the chemical and volatile composition of sawdust smoke and friction smoke are compared. It can be seen that both the steam volatile and nonsteam volatile fractions of friction smoke are more acidic both in pH and in percent total acids. In addition, friction smoke contained four more steam volatile acids and approximately eight times more carbonyls than sawdust smoke. Overall, friction smoke contained more steam volatiles that are known to be important to smoke flavor than sawdust smoke. However, these differences may be due in part to the fact that friction smoke was concentrated, since it only contained 71% moisture, whereas sawdust smoke contained 87% moisture.

Even more dramatic is the comparison of hot vs. cold smoking of a product relative to the number and amount of phenols present. Data of these type are seen in Table 43. As you can see, guaiacol, maltol, and phenol were only found in measurable quantities in the cold smoking process which had approximately three times more eugenol than the hot process. In contrast, the hot process contained major amounts of acetovanillone and acetosyringone, which were hardly present in the cold process. Differences as dramatic as this undoubtedly account for significant differences in resulting product-sensory properties. Thus, it becomes very obvious that the smoking process employed can play a major role in influencing the resulting sensory properties of the product involved.

VI. STORAGE

In the generation and utilization of liquid smoke, research has demonstrated that chemical changes that occur with time can alter the sensory properties of smoke. This has potential significance with the long-term storage of liquid smoke. This fact is summarized in Table 44. As can be seen, the newest smoke (5 hr old) had the lowest sensory score, but with time, the score gradually improved. They postulated that with time, methyl alcohol can react with formic and acetic acids to form less harsh-flavored esters.

Doerr et al.[14] actually followed the formation of methyl acetate and methyl formate in liquid smoke stored at 25°C over a period of approximately 30 days, and as seen in Table 45, the amount of methyl formate formed reached equilibrium in approximately 70 hr, whereas the level of methyl acetate was still gradually increasing after 700 hr. Undoubtedly, other chemical changes can occur during liquid smoke storage that can either result in the loss of desirable or undesirable compounds with the formation of other flavor-related compounds. More study in this area is also encouraged, especially with long-term storage.

Table 1
COMPOUNDS IDENTIFIED IN WOOD SMOKE

Acids	Ref.
Formic (methanoic)	18,22,33,36,48
Acetic (ethanoic)	4,6,12,16,18,22,24,25,29,32,33,36,37,39,43,48
Glycolic (hydroxyethanoic)	48
Propionic (propanoic)	4,6,12,16,18,20,22,24,29,32,33,36,37,43,48
Isobutyric (2-methylpropanoic)	4,18,20,29,32,33,37,48
Acrylic (propenoic)	48
Methacrylic (methylpropenoic)	20,48
Butyric (butanoic)	4,6,16,18,20,22,24,29,32,33,36,37,48
α-Methylbutyric	48
Isovaleric (3-methylbutanoic)	4,16,18,20,29,32,33,37,38
Crotonic (2-butenoic)	33,36,48
trans Crotonic (*trans* 2-butenoic)	4,29,32,37
cis Crotonic (*cis* 2-butenoic)	4,29,37
3-Butenoic	48
Tiglic (2-methyl-2-butenoic)	1,6,33,36,48
Angelic (*cis* 2-methyl-2-butenoic)	20,37,48
trans 2-methyl-2-butenoic	4,20,32,37
Valeric (pentanoic)	4,6,16,18,20,24,29,32,33,36,37,48
α-Methylvaleric (2-methylpentanoic)	48
3-Methylpentanoic	20
Isocaproic (4-methylpentanoic)	16,20,37,48
4-Methyl-2-pentenoic	48
2-Pentenoic	4,48
3-Pentenoic	48
Levulinic (4-oxopentanoic)	6,48
Caproic (hexanoic)	16,18,20,24,32,37,48
Enanthic (heptanoic)	18,32,33,37,48
5-Heptenoic	48
Caprylic (octanoic)	18,20,32,37,48
Dehydroxycaprylic	48
Pelargonic (nonanoic)	18,32,48
Capric (decanoic)	18,32,48
Lauric (dedecanoic)	37
Myristic (tetradecanoic)	32,34,37
Pentacecylic (pentadecanoic)	32
Palmitic (hexadecanoic)	32,34,37,48
Stearic (octadecanoic)	48
Oleic (9-octadecanoic)	48
Arachidic (eicosanoic)	48
Behinic (docosanoic)	48
Benzoic	20,33,37
Salicylic (*O*-hydroxybenzoic)	30,46
3-Methoxy-4-hydroxybenzoic	44
Sorbic (2-4-hexadienoic)	33
Vanillic (4-hydroxy-3-methoxybenzoic)	41
Abietinic	48
Pimaric	48
Lignoceric	48
Alcohols	
Methyl (methanol)	4,13,14,36,39,48
Ethyl (ethanol)	4,13,14,20,39,43,46,48
Propyl (1-propanol)	48
Isopropyl (2-propanol)	14,39,48

Table 1 (continued)
COMPOUNDS IDENTIFIED IN WOOD SMOKE

	Ref.
Isobutyl (2-methyl-1-propanol)	48
Allyl (2-propen-1-ol)	4,13,48
Propan-2-on-1-ol	4
Amyl (2-methyl-2-butanol)	4
Isoamyl (3-methyl-2-butanol)	34,48
Butan-2-on-1-ol	4
Butan-3-on-2-ol	4
2-Buten-1-ol	48
Pentan-2-on-1-ol	4
Pentan-3-on-2-ol	4
1-Penten-3-ol	35
1-Nonanol	30
Phenylethylalcohol	4,34
Cyclohexanol	48
Benzylalcohol	4,43
p-Hydroxybenzylalcohol	46
Fenchyl (6-fenchanol)	48
Isofenchyl (iso-6-fenchanol)	48
Carbonyls	
Methanal	13,15,18,48
Ethanal	13—15,18,20,36,39,48
Hydroxyethanal	48
Phenylethanal	34,35
2-Pyrroaldehyde	43
Propanal	13—15,18,20,48
2-Propanal (Acrolein)	15
2-Methylpropanal	13,15
2,2-Dimethylpropanal	48
2-Methylpropenal	13
Acetone (2-propanone)	4,13—15,18,20,39,48
Acetylacetone	30
3-Methylacetylacetone	30
1-Hydroxypropanone (acetol)	6,14,24,25,30,39,48
2-Hydroxypropanone	30
Acetoxypropanone	48
1,3′-Dimethoxy-4-phenyl-1-propanone	30
1,3′-Dimethoxy-4-phenyl-2-propanone	30
1,5′-Dimethoxy-4-phenyl-2-propanone	30
α-Methyl-β-ethylpropenal	48
Butanal	13,15,18
2-Butenal	13—15,18,39,48
3-Methylbutanal	14,15,18,34,39,48
2-Methyl-2-butenal	15,20,46,48
2-Butanone	4,13,15,18,46
3-Methyl-2-butanone	15,28
1-Hydroxy-2-butanone	6
3-Hydroxy-2-butanone	30
1-Phenyl-2-butanone	30
3,3-Dimethyl-2-butanone	15
Diacetyl (2,3-butanedione)	4,13—15,20,30,36,39,48
Pentanal	13,15,18,48
α-Methylpentanal	15
2-Methyl-2-pentenal	43
2-Pentanone	4,15,18,30,48

Table 1 (continued)
COMPOUNDS IDENTIFIED IN WOOD SMOKE

Carbonyls	Ref.
3-Pentanone	15,48
4-Hydroxy-2-pentanone	48
4-Hydroxy-4-methyl-2-pentanone	37
2-Methyl-3-pentanone	48
2,4-Dimethylpentanone	30
Pentadiene	46
Methylpentadiene	46
2,3-Pentanedione	20,35,48
4-Pentanedienal	30
2,4-Pentanedienal	43
Hexanal	35
2-Hexenal	35
2-Hexanone	15,30
3-Hexanone	15,48
5-Methyl-2-hexanone	30
2-Methyl-3-hexanone	30
3-Hexen-2-one	48
5-Hexen-2-one	30
2,4-Hexadienal	37
2,4-Hexanedione	4
2,5-Hexanedione	6,30,43
3-Methyl-2,5-hexanedione	30
2-Heptanone	30
4-Heptanone	30,37,48
3,6-Octanedione	48
Methyl ethyl ketone	20,28,30
Methyl isopropyl ketone	30,43
Methyl isopropenyl ketone	20
Methyl butyl ketone	43
Methyl vinyl ketone	14,39
Butyl methyl ketone	48
Allyl methyl ketone	4
1-Cyclohexyl methyl ketone	37
Cyclic ketone	4
Acetophenone (methyl phenyl ketone)	4,20,43
m-Methylacetophenone	4
p-Methylacetophenone	4
Acetosyringone(4′-hydroxy-3′,5′-dimethoxyacetophenone)	21,23,26,27,45
Propiosyringone(4′hydroxy-3′,5′dimethoxypropioacetophenone)	21,23
4′-Hydroxy-3′-methoxyacetophenone	23
Benzaldehyde	4,20,30
Salicyaldehyde (2-hydroxybenzaldehyde)	21,46
Anisaldehyde (4-hydroxybenzaldehyde)	30,43,45,46
Dihydroxybenzaldehyde	45
1,2-Dihydroxy-3-methylbenzaldehyde	42
2-Hydroxy-3-methoxybenzaldehyde	30
4-Hydroxy-2-methoxybenzaldehyde	30
4-Hydroxy-3-methoxybenzaldehyde (vanillin)	3,5,6,21,23—27,30,37,39,41,43—46
4-Hydroxy-3,5-dimethoxybenzaldehyde (syringaldehyde)	3,11,21,23,26,27,30,44—46
3,5-Dimethoxybenzaldehyde	44
Sinapaldehyde (3-(4′-hydroxy-3′,5′dimethoxyphenyl)-2-propenal)	11,21,23,25
Coniferaldyhyde (3-(4′-Hydroxy-3′-methoxyphenyl)-2-propenal)	21,23

Table 1 (continued)
COMPOUNDS IDENTIFIED IN WOOD SMOKE

Carbonyls	Ref.
3-(4′-Hydroxy-3′,5′-dimethoxyphenyl)-1-propanal	45
3-(4′-Hydroxy-3′-methoxyphenyl)-l-propanal	45
Cyclopentanone	4,29,30,43,48
Methylcyclopentanone	4,48
Dimethylcyclopentanone	48
Trimethylcyclopentanone	48
3,3,5-Trimethylcyclopentanone	48
1,2-Cyclopentanediol	48
3-Methyl-1,2,4-cyclopentanetrione	37
3-Methylcyclopenta-1,2-dione	4
3-Ethylcyclopenta-1,2-dione	4
3,4-Dimethylcyclopenta-1,2-dione	4
2-Cyclopentenone	4,25,39,43,46,48
2-Methyl-2-cyclopentenone	4,6,29,48
4-Methyl-2-cyclopentenone	4,48
5-Methyl-2-cyclopentenone	4
2-Ethyl-l-cyclopentenone	4
3-Ethyl-2-cyclopentenone	4
2-Ethyl-3-methyl-2-cyclopentenone	4
2-Ethyl-4-methyl-2-cyclopentenone	4
2-Ethyl-5-methyl-2-cyclopentenone	4
3-Ethyl-2-methyl-2-cyclopentenone	4
2,3-Dimethyl-2-cyclopentenone	4,6,29,43
2,4-Dimethyl-2-cyclopentenone	4
2,5-Dimethyl-2-cyclopentenone	4
3,4-Dimethyl-2-cyclopentenone	4
3,5-Dimethyl-2-cyclopentenone	4
2,3,4-Trimethyl-2-cyclopentenone	4
2,3,5-Trimethyl-2-cyclopentenone	4
3-*n*-Propyl-2-cyclopentenone	4
3-Isopropyl-2-cyclopentenone	48
3-Methyl-2-cyclopenten-2-ol-1-one	1,4,20,29,43,44
3,4-Dimethyl-2-cyclopenten-2-ol-1-one	1,4,6
3-Ethyl-2-cyclopenten-2-ol-1-one	1,4,6
2-Hydroxy-3-methylcyclopent-2-ene-1-one (cyclotene)	2,3,5,6,8,21,23,25,26,30,37,39,43,46,48
Cyclohexanone	4,43,48
2,4,4-Trimethylcyclohexanone	48
2-Cyclohexenone	20,43
2,4-Dimethyl-4-cyclohexenone	48
1-Methyl-2-cyclohexen-5-one	48
5,5-Dimethyl-1,3-cyclohexanedione	30
Cyclohexanecarboxaldehyde	48
Cyclohexanepropionaldehyde	48
Esters	
Methyl formate	13,14,20,39,48
Methyl acetate	13,14,39,48
Methyl propionate	48
Methyl butyrate	4,48
Methyl isobutyrate	48
Methyl crotonate	48
Methyl valerate	48
Methyl caproate	48
Methyl enanthate	48
Methyl myristate	34,35
Methyl palmitate	35

Table 1 (continued)
COMPOUNDS IDENTIFIED IN WOOD SMOKE

Esters	Ref.
Methyl acrylate	20
Ethyl myristate	34
Ethyl benzoate	20
Cresyl acetate	20
Hydroxy-2-butanone acetate	4,6,48
Hydroxy-2-propanone propionate	48
Salicylic acid methyl ester	46
4-Methoxybenzoic acid methyl ester	43
4-Hydroxy-4-methoxybenzoic acid methyl ester	45
4-Hydroxy-3,5-dimethoxybenzoic acid methyl ester	45
3-Methoxy-4-hydroxyphenylacetic acid methyl ester	45
Furans	
Furfuryl alcohol	2,21,25,30,33—37,39,43,46,48
2-Furfuryl methyl alcohol	37
Furan	13,14,20,39,46,48
2-Methylfuran	20,36,39,46
2,5-Dimethylfuran	46,48
2,4-Dimethylfuran	43
2-Ethylfuran	17,20
2-Acetylfuran	17,20,29,43,46
Propylfuran	48
2-Isopropylfuran	43
2-Propenylfuran	43
2-Acetyl-5-methylfuran	43
2,5-Dimethyl-3-acetylfuran	43
Amylfuran	48
2,3,5-Trimethylfuran	48
2-Vinylfuran	46
2-Vinyl-3-methylfuran	46
Benzofuran	43,46,48
2-Methylbenzofuran	20,37,43,46
Dimethylbenzofuran	20,37
Dibenzofuran	46
Bis-(furfuryl)-2-furan	43
2,3-Dimethyl-5-ethylfuran	17
3,4-Dimethyl-5-ethylfuran	17
2-Ethyl-2,3-dihydro-5-methylfuran	48
2,5-Tetrahydrodimethylfuran	48
2-Methyltetrahydrofuran-3-one	4
2,5-Dimethyltetrahydrofuran-3-one	1,4,6
2-Acetyltetrahydrofuran-3-one	4
2-Furoic acid	48
4-Methyl-2-furoic acid	48
Methyl-2-furoate	4,48
Furfuryl methyl ether	4,37
2-Furfuryl methyl ketone	4,6,37,48
3-Furfuryl methyl ketone	37
2-Furfuryl propyl ketone	30
2-Furfuryl butyl ketone	30
2-(5-Methylfuryl)-methyl ketone	4
2-Furfural	4,6,12,20,21,24,25,29,30,33,35,36,39,43,46,48
3-Furfural	30,46
4-Methylfurfural	17,30
5-Methylfurfural	4,6,12,14,17,20,21,24,25,29,30,33—35,39,43,46,48

Table 1 (continued)
COMPOUNDS IDENTIFIED IN WOOD SMOKE

	Ref.
2-Methyl-3-furfural	30,48
3-Methyl-2-furfural	30,48
5-Hydroxymethyl-2-furfural	6,43,48
2,5-Difurfuryledine-1-cyclopentanone	48
Lactones	
γ-Butyrolactone	4,6,48
β-Angelica lactone	6
γ-Crotonolactone	6
γ-Hydroxyvalerolactone	48
δ-Hydxroxyvalerolactone	48
α-Hydroxy-γ-hydroxyvalerolactone	48
2-Butenolide	4
2-Methyl-2-butenolide	4
4-Methyl-2-butenolide	4
2,3-Dimethyl-2-butenolide	4
2,4-Dimethyl-2-butenolide	4
3,4-Dimethyl-2-butenolide	4
2,3,4-Trimethyl-2-butenolide	4
4-Ethyl-2-methyl-2-butenolide	4
4-Ethylidiene-2-methyl-2-butenolide	4
Methylvinyl-2-butenolide	4
Phenols	
Phenol	1—6,10,12,17,19—21,23,25—27,29,32—35,37—40,42—46,48
o-Cresol (2-methylphenol)	4,6,12,17,20,21,23,27—29,32—35,37,38,40,43—48
m-Cresol (2-methylphenol)	4—6,10,12,17,20,21,23,25—27,29,32—35,37,38,40,42—46,48
p-Cresol (4-methylphenol)	4—6,10,12,17,20,21,23,26,27,29,32—35,37,38,40,43—46,48
2-Ethylphenol	20,23,33,43,45,46,48
3-Ethylphenol	20,27,43,45
4-Ethylphenol	1,6,20,27,42—46
2-Ethyl-5-methylphenol	4
3-Ethyl-5-methylphenol	4,20
2,3-Xylenol (2,3-dimethylphenol)	4,20,23,32—35,37,43—45,48
2,4-Xylenol (2,4-dimethylphenol)	1,4,6,20,23,29,32—35,43,45,46,48
2,5-Xylenol (2,5-dimethylphenol)	6,20,37,43,45
2,6-Xylenol (2,6-dimethylphenol)	4,11,20,23,27,43,45,46
3,4-Xylenol (3,4-dimethylphenol)	1,6,7,17,23,39,43—46
3,5-Xylenol (3,5-dimethylphenol)	4,12,23,29,37,44,45,48
Diethylphenol	43,44,46
2,3,5-Trimethylphenol	4,20,23,27,37,43,45,46
2,3,6-Trimethylphenol	45
2,4,6-Trimethylphenol	44—46
4-Propylphenol	27,45
4-Isopropylphenol	33,43,44,46
2-Methyl-5-isopropylphenol	43,46
4-Butylphenol	20,37
2-Isobutylphenol	46
2-Allylphenol	45
3-Allylphenol	45
4-Allylphenol	45
4-Vinylphenol	23,44,45

Table 1 (continued)
COMPOUNDS IDENTIFIED IN WOOD SMOKE

Phenols	Ref.
3-Methoxyphenol	20
4-Methoxyphenol	44
Dimethoxyphenol	45
Methoxyethylphenol	45
3,4-*Dimethoxyphenol*	43
Syringol (2,6-dimethoxyphenol)	3—6,9,12,17,20,21,23—25,27—29,31—34,37—40,42—46,48
4-Methylsyringol (2,6-dimethoxy-4-ethylphenol)	3,4,6,9,12,17,20,21,23—25,27,29,31,32,34,35,37,39,40,42—46,48
4-Ethylsyringol (2-dimethoxy-4-ethylphenol)	3,4,6,9,11,17,20,21,23,25,29,32,34,35—37,39,40,42—46,48
4-Propylsyringol (2,6-dimethoxy-4-propylphenol)	4,6,20,23,25,37,39,40,44—46
4-Isopropylsyringol (2,6-dimethoxy-4-isopropylphenol)	20
4-Propenylsyringol (2,6-dimethoxy-4-propenylphenol)	4,20,38,42,43,46
4-*trans* Propenylsyringol (2,6-dimethoxy-4-*trans* propenylphenol	21,23,45
4-*cis* Propneylsyringol (2,6-dimethoxy-4-*cis* propenylphenol)	21,23,44,45
4-Allysyringol (2,6-dimethoxy-4-allylphenol)	5,8,17,20,21,23,24,26,27,37,38,42,44—46
4-Vinylsyringol (2,6-dimethoxy-4-vinylphenol)	21,23,46
4-Isobutylsyringol (2,6-dimethoxy-4-isobutylphenol)	44
2,4-Dimethoxy-4-allylphenol	42
1,2-Dihydroxy-4-methylphenol	42
2-Methoxy-4-acetic acid phenol	44
3-Methoxy-4-propan-2-one phenol	44
Acetophenol	37
4-Isopropylacetophenol	37
4-Hydroxy-3-methoxyacetophenol	37
3,4-Dimethoxycetophenone	44
2-Acetovanillone (4′-hydroxy-3′-methoxyacetophenone)	4—6,11,21,26,27,33,41,45,46
Guaiacol (2-methoxyphenol)	1—4,6,10,12,19—21,23,25—27,29,31—35,37—48
4-Methylguaiacol (2-methoxy-4-methylphenol)	3—6,10,11,17,19—21,23—25,27,29,31—35,37,39—47
6-Methylguaiacol (2-methoxy-6-methylphenol)	45,48
5-Methylguaiacol (2-methoxy-5-methylphenol)	45
3-Methylguaiacol (2-methoxy-3-methylphenol)	4,27,45
4-Ethylguaiacol (2-methoxy-4-ethylphenol)	4—7,17,20,21,23,25,29,31—35,37,39—46,48
4-Propylguaiacol (2-methoxy-4-propylphenol)	4,6,20,23,25,32—35,37,39,40,43—45,48
4-Isopropylguaiacol (2-methoxy-4-isopropylphenol)	20
4-Vinylguaiacol (2-methoxy-4-vinylphenol)	4,6,20,21,23,25,32—35,37,39,41,44—46,48
5-Vinylguaiacol (2-methoxy-5-vinylphenol)	7,17
3-5-Dimethylguaiacol (2-methoxy-3,5-dimethylphenol)	48
Eugenol (2-methoxy-4-allylphenol)	4,6,7,17,20,21,23,25—27,32—34,37,49,40,42—46
trans-Isoeugenol (2-methoxy-4-*trans*-propenylphenol)	3,5,6,11,20,21,23,24,26,27,32—35,37,40,42—46,48
cis-Isoeugenol (2-methoxy-4-*cis*-propenylphenol)	5,6,21,23,27,37,43,44,45
Pyrocatechol (2,3-benzenediol)	4,37,44
3-Methylpyroacatechol	4,27,37,45
4-Methylpyrocatechol	4,27,37,44
5-Methylpyrocatechol	44
4-Ethylpyrocatechol	4
3-Methoxypyrocatechol	4,6,27,45
Resorcinol (1,3-dihyroxybenzene)	20,46
4-Ethylresorcinol	27,45

Table 1 (continued)
COMPOUNDS IDENTIFIED IN WOOD SMOKE

Miscellaneous	Ref.
Methylamine	48
Dimethylamine	48
Trimethylamine	48
Pyrazine	4
Methylpyrazine	4
2,5-Dimethylpyrazine	4
Pyrrole	33
2-Acetylpyrrole	43,46
2-Formlypyrrole	4
α-Methyl pyrryl ketone	4
α-Ethyl pyrryl ketone	4
Pyridine	48
4-Hydroxy-6-methyl-2,4-pyran-2-one	37
Maltol (3-hydroxy-2-methylpyrone)	1,3,6,21,23,36,30,33,37,45,48
Dimethoxymethane	48
Dimethoxyethane	48
Methoxybenzene	43
Thymol	43
Ethanediol	48
1-Naphthol	44
Toluene	13,20
Styrene	20
Benzene	13,20,48
Ethylbenzene	4
5-Propyl-1,2,3-benzentriol	48
1,2-Dimethoxybenzene (veratrol)	4,20,25,39,43
1,2-Dimethoxy-4-methylbenzene (methylveratrol)	4,20,43,48
1,2-Dimethoxy-4-ethylbenzene (ethylveratrol)	4
1,4-Dimethoxybenzene	20,43
n-Pentadecane	32
2,6,10,14-Tetramethylpentadecane	32,33
1-Indanone	4,30,43,46
2-Methyl-1-indanone	4
3-Methyl-1-indanone	4
6-Methyl-1-indanone	4
Indene	20,37,46
3-Methylindene	20,46
Trimethylindene	46
2,3-Dihydroindene	46
Dimethylindene	20,46
Naphthalene	20,46,48
1-Methylnaphthalene	46
2-Methylnapthalene	20,46
2-Ethylnaphthalene	46
Trimethylnaphthalene	46
Methylethylnaphthalene	46
Methylvinylnaphthalene	46
Allylnaphthalene	44
6-Methyl-1,2-dihydronaphthalene	46
Methyldihydronaphthalene	46

Table 2
TYPICAL FRACTIONATION SCHEME FOR WOOD SMOKE VOLATILES

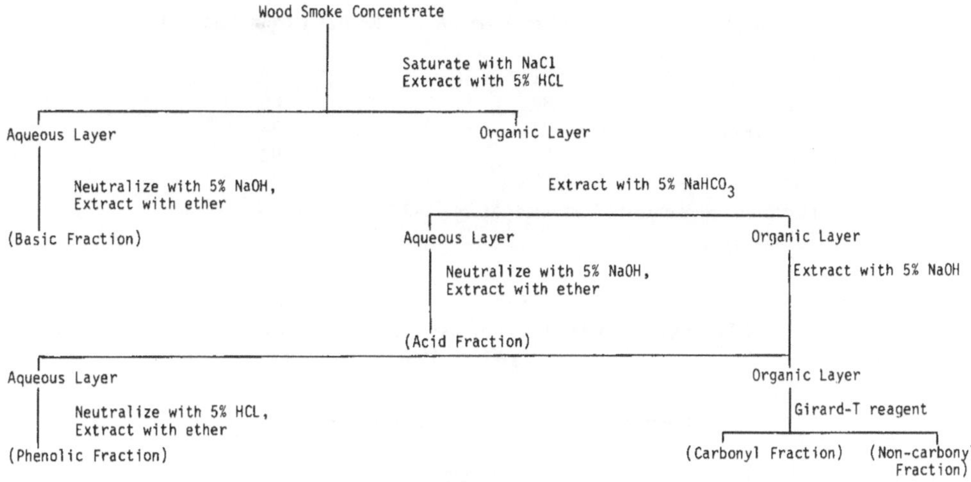

Table 3
CONCENTRATIONS OF MAJOR PHENOLS IN WOOD SMOKE

Compound	Whole smoke (mg/ℓ)	Vapor phase (mg/ℓ)
Phenol	59	5.9
Guaiacol	417	32.1
4-Methylguaiacol	333	13.6
Syringol	392	6.5
Total	1201	58.1

Adapted from Kornreich, M. R. and Issenberg, P., *J. Agric. Food Chem.*, 20, 1109, 1972.

Table 4
% LOSS OF PHENOLS WITH 40 DAYS STORAGE AT −4°C FROM SMOKED COD

Compound	% Remaining
Vanillin	0
2,3-Dimethylphenol	0
Eugenol	0
4-Ethylguaiacol	13
3,4-Dimethylphenol	24
4-Methylguaiacol	34
4-Ethylphenol	44
Isoeugenol	51
3,5-Dimethylphenol	54
Guaiacol	55
2,6-Dimethoxyphenol	63
Phenol	67
m-/p-Cresol	69

Adapted from McWeeny, D. J., *Food Chem.*, 9, 89, 1982.

Table 5
TOTAL PHENOL (mg/100 g) DISTRIBUTION IN A SMOKED BOLOGNA EMULSION

Layer	Concentration
Outer 1.6 mm	3.70
Next 1.6 mm	2.04
Next 1.6 mm	1.41
Next 1.6 mm	1.02
Next 1.6 mm	0.78
Next 1.6 mm	0.43
Next 1.6 mm	0.26
Inner 1.6 mm	0.12
Raw emulsion	0.00

Adpated from Bratzler, L. J., Spooner, M. E., Weatherspoon, J. B., and Maxey, J. A., *J. Food Sci.*, 34, 146, 1969.

Table 6
INFLUENCE OF MEDIA ON PHENOL RECOVERY

Compound	Recovery from water (%)	Recovery from lipid (%)
Phenol	81	89
Guaiacol	80	80
Methylguaiacol	77	84
Syringol	99	92

Adapted from Issenberg, P., Kornreich, M. R., and Lustre, A. O., *J. Food Sci.*, 36, 107, 1971.

Table 7
PHENOLICS IDENTIFIED IN CHARRED OAK

Phenols	Phenolic acids
Phenol	Benzoic acid
o-Cresol	*p*-Hydroxybenzoic acid
m-Cresol	2,3-Dihydroxybenzoic acid
p-Cresol	3,5-Dihydroxybenzoic acid
p-Cresol	2,6-Dihydroxybenzoic acid
Guaiacol	2,3,4-Trihydroxybenzoic acid
2,4-Dimethylphenol	3-Methoxy-4-hydroxybenzoic
2,6-Dimethylphenol	2-Methoxy-4-hydroxybenzoic
2,6-Dimethoxy-4-methylphenol	2,3-Dimethoxybenzoic acid
2,6-Dimethoxy-4-ethylphenol	3,4-Dihydroxycinnamic acid
Acetoxyringone	Phenylacetic acid
Eugenol	Phenylbutric acid
p-Hydroxybenzyl alcohol	*m*-Hydroxyphenylacetic acid

Adapted from Maga, J. A., *Progress in Flavour Research*, Adda, J., Ed., Elsevier, Amsterdam, 1985, 409.

Table 8
THE FLAVOR THRESHOLD (ppm) AND FLAVOR INDEX[a] IN WATER AND OIL OF SEVERAL IMPORTANT SMOKE ASSOCIATED PHENOLS

	Water				Oil	
Compound	Taste threshold	Taste index	Odor threshold	Odor index	Odor threshold	Odor index
Guaiacol	0.013	6,400	0.021	4,600	0.07	1,000
4-Methylguaiacol	0.065	90,000	0.09	58,800	0.40	18,000
2,6-Dimethoxyphenol	1.65	1,400	1.85	1,200	0.34	7,000

[a] Concentration in smoke condensate ÷ mean odor threshold concentration.

Adapted from Wasserman, A. E., *J. Food Sci.*, 31, 1005, 1966.

Table 9
SENSORY CHARACTERIZATION OF SEVERAL PHENOLS ASSOCIATED WITH WOOD SMOKE

	% characterizing compound as					
	Taste			Odor		
Compound	Smoky	Phenolic	Bitter	Smoky	Phenolic	Sweet
2,6-Dimethoxyphenol	46.5	34.0	17.0	71.5	31.5	—
Guaiacol	58.6	33.5	9.5	53.5	37.5	5.2
4-Methylguaiacol	43.2	36.3	26.0	51.8	35.5	13.8

Adapted from Wasserman, A. E., *J. Food Sci.*, 31, 1005, 1966.

Table 10
ODOR DESCRIPTIONS OF VARIOUS PHENOLS ISOLATED FROM WOOD SMOKE

Compounds	Description
Phenol	Pungent
o-Cresol	Pungent
m- and *p*-Cresol	Pungent
2,3-Xylenol	Pungent
2,4-Xylenol	Pungent
2,6-Xylenol	Cresolic
3,4-Xylenol	Cresolic
3,5-Xylenol	Cresolic
2-Ethyl-5-methylphenol	Cresolic
3-Ethyl-5-methylphenol	Cresolic
2,3,5-Trimethylphenol	Cresolic
Guaiacol	Sweet, smoky, somewhat pungent
3-Methylguaiacol	Weak, phenolic
4-Methylguaiacol	Sweet, smoky
4-Ethylguaiacol	Sweet, smoky
4-Allyguaiacol	Woody
2,6-Dimethoxyphenol	Smoky
2,6-Dimethoxy-4-methylphenol	Mild, heavy, burnt
2,6-Dimethoxy-4-ethylphenol	Mild, heavy, burnt
2,6-Dimethoxy-4-propylphenol	Mild, heavy, burnt
4,6-Dimethoxy-4-propenylphenol	Mild, heavy, burnt
Pyrocatechol	Heavy, sweet, burnt
3-Methylpyrocatechol	Heavy, sweet, burnt
4-Methylpyrocatechol	Heavy, sweet, burnt
4-Ethylpyrocatechol	Heavy, sweet, burnt

Adapted from Kim, K., Kurata, T., and Fujimaki, M., *Agric. Biol. Chem.*, 38, 53, 1974.

Table 11
SENSORY DESCRIPTIONS OF VARIOUS SMOKE-ASSOCIATED PHENOLS

Compound	Optimum sensory concentration (mg/100 mℓ)	Odor description	Flavor description
Dimethylphenol	0.9	Phenolic, like ink, aromatic, sweet	Phenolic, sharp, charred sweet, dry
4-Methylguaiacol	1.9	Sweet, like vanilla, fruity, like cinnamon, somewhat smoky, pleasantly sharp, phenol tones	Sweet, like vanilla, caramel-like, aromatic, pleasant smoke tones, burning
Guaiacol	3.75	Phenolic, smoky, aromatic, sharp, sweet	Phenolic, sharp, spicy smoked-sausage-aromatic, sweet, dry
Syringol	7.50	Smoky, spicy, aromatic, smoked-sausage, phenolic, sharp, sweet	Phenolic, smoky, freshly charred wood, like whiskey, dry, sharp
o-Cresol	7.5	Phenolic, sweet-fruity, aromatic, like caramel, smoked-sausage	Sweet, sharp, unpleasant smoky, burning
Isoeugenol	9.8	Sweet-fruity, like vanilla, like rhubarb, phenolic	Sweet-fruity, mild smoke, dry, sharp

Adapted from Baltes, W., Wittkowski, R., Sochtig, I., Block, H., and Toth, L., *The Quality of Food and Beverages*, Vol. 2, Charalambous, G. and Inglett, G., Eds., Academic Press, New York, 1981, 1.

Table 12
SENSORY PROPERTIES OF INDIVIDUAL PHENOLS

Phenol	Optimum concentration (mg/100 mℓ H_2O)	Sensory description	
		Odor	Taste
Dimethylphenol	0.90	Phenolic, ink, aromatic, sweet	Phenolic, hot, sweet, dry
4-Methylguaiacol	1.90	Sweet, vanilla-like, fruity, like cinnamon, smoky	Sweet, vanilla-like, caramel-like, aromatic, pleasant smoke
Cyclotene	1.90	Like hydrolyzed vegetable protein, smoked ham, spicy, aromatic, like clove	Like hydroxyzed vegetable protein, like smoked ham, sweet, hot
Guaiacol	3.75	Phenolic, smoky, aromatic, hot, sweet	Phenolic, hot spicy, smoked ham, sweet, dry
o-Cresol	7.50	Phenolic, sweet, fruity, aromatic, caramel-like, like smoked ham	Sweet, hot, unpleasant, smoky, burning
Isoeugenol	9.80	Like hydrolyzed vegetable protein, ham, spicy, aromatic like clove	Like hydrolyzed vegetable protein, like smoked ham, sweet, hot

Adapted from Toth, L. and Potthast, K., *Adv. Food Res.*, 29, 87, 1984.

Table 13
EFFECT OF SMOKE FRACTION ADDITIONS ON RESULTING AROMA PROPERTIES

Fraction additions	Aroma description	Aroma quality
Phenolic only	Montonous, smoky	Good
Phenolic + carbonyl	Smoky, with fragrance	Better
Phenolic + carbonyl + noncarbonyl	Smoky, with fragrance	Best
Phenolic + neutral + basic	Burnt, smoky	Best

Adapted from Fujimaki, M., Kim, K., and Kurata, T., *Agric. Biol. Chem.*, 38, 45, 1974.

Table 14
SENSORY INTENSITIES OF LIQUID SMOKE FRACTIONS

	Fraction					
Taste property	**1**	**2**	**3**	**4**	**5**	**6**
Smoke taste intensity	6	7	3	11	4	10
Tarry taste intensity	3	1	2	0	6	1
Chemical taste intensity	1	1	3	0	1	0
Acidulous taste intensity	1	2	3	0	0	0

Note: Intensity scale: 0 = below threshold; 11 = highest value; (1) distilled at 67 to 90°C; (2) distilled at 91 to 132°C; (3) distilled at 133 to 200°C; (4) phenol subfraction; (5) terpene subfraction; and (6) whole liquid smoke.

Adapted from Olsen, C. Z., *Proc. Eur. Meet. Meat Res. Workers*, No. 22, F7:1, 1976.

Table 15
ODOR AND TASTE RECOGNITION THRESHOLDS (ppm) AND MOST DESIRABLE CONCENTRATIONS (ppm) OF THE PHENOLIC FRACTION ISOLATED FROM THE VAPOR AND PARTICULATE PHASES OF WOOD SMOKE

Odor

Recognition threshold		Most desirable concentration	
Vapor	Particulate	Vapor	Particulate
10.4	7.8	20.8	16.7

Taste

Recognition threshold		Most desirable concentration	
Vapor	Particulate	Vapor	Particulate
2.3	1.4	15.6	8.3

Adapated from Daun, H., *Lebensm. Wiss. Technol.*, 5, 102, 1972.

Table 16
CONVERSION OF FERULIC ACID TO 4-VINYLGUAIACOL

Adapted from Fiddler, W., Parker, W. E., Wasserman, A. E., and Doerr, R. C., *J. Agric. Food Chem.*, 15, 757, 1967.

Table 17
THERMAL DECOMPOSITION PRODUCTS OF FERULIC ACID

Guaiacol
4-Methylguaiacol
4-Ethylguaiacol
4-Vinylguaiacol
cis-Isoeugenol
Vanillin
Acetovanillone

Adapted from Fiddler, W., Parker, W. E., Wasserman, A. E., and Doerr, R. C., *J. Agric. Food Chem.*, 15, 757, 1967.

Table 18
CONVERSION OF 4-VINYLGUAIACOL TO VARIOUS BY-PRODUCTS

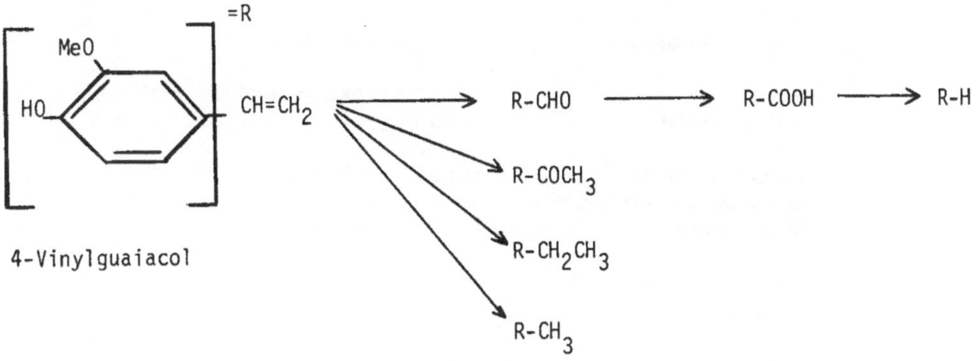

Adapted from Fiddler, W., Parker, W. E., Wasserman, A. E., and Doerr, R. C., *J. Agric. Food Chem.*, 15, 757, 1967.

Table 19
PHENOLS ASSOCIATED WITH VARIOUS SPICES

Spice	Phenols
Cinnamon	Phenol
	O-Cresol
	2-Vinylphenol
	4-Allylphenol
	Guaiacol
	4-Ethyl-2-methoxyphenol
	Eugenol
	Methyleugenol
	Acetyleugenol
Thyme	2-Isopropyl-5-methylphenol
	5-Isopropyl-2-methylphenol
Pepper	Eugenol
	Methyleugenol
Nutmeg	Eugenol
	Methyleugenol
	Methoxyeugenol
	Isoeugenol
	Methylisoeugenol
Marjoram	5-Isopropyl-2-methylphenol
	4-Allylphenol
	Eugenol
Cloves	4-Allylphenol
	Eugenol
	Acetoeugenol

Adapted from Baltes, W. and Sochtig, I., *Z. Lebensm. Unters. Fursch.*, 169, 17, 1979.

Table 20
ODOR DESCRIPTIONS OF CERTAIN FURANS ISOLATED FROM WOOD SMOKE

Compound	Odor description
Furfural	Sweet, bread-like, caramel-like
5-Methylfurfural	Sweet-spicy, warm, slightly caramel-like
2-Furylmethylketone	Powerful, balsamic, sweet
2-(5-Methylfuryl)-methylketone	Sweet, fragrant
Methyl furoate	Grassy, acetophenone-like

Adapted from Kim, K., Kurata, T., and Fujimaki, M., *Agric. Biol. Chem.*, 38, 53, 1974.

Table 21
VOLATILES IDENTIFIED IN THE BASIC FRACTION OF FIRE-CHARRED WHITE OAK

Pyrazines

Pyrazine
2-Methyl-3-isopropylpyrazine
2-Methoxy-3-ethylpyrazine
2-Methoxy-3-methylpyrazine
2-Methoxy-3-propylpyrazine
2-Acetylpyrazine
2-Acetyl-3-methylpyrazine
2-Acetyl-3,5-dimethylpyrazine
2-Methyl-5-acetylfuran

Furans

2,5-Dihydrofuran
3-Methyl-2,3-dihydrofuran
3-Methyl-2,5-dihydrofuran
3-Methyl-2(5H)-furanone
5-Methyl-2(5H)-furanone
2,5-Dimethylfuran
3,4-Dimethylfuran
2,3,4-Trimethylfuran
2,5-Diethylfuran
2-Furancarboxaldehyde
2-Methyl-5-ethylfuran
2-Furanylethanone

Miscellaneous

3,5-Dimethylpyrazole
Dimethylmelamine
2-Isopropylfuran
4-Propenyl-2,3-dihydrofuran
2-Methyl-5-isopropylfuran
2-Propyl-5-methylfuran
2-Furfuryl alcohol
5-Methyl-2-oxo-2,3-dihydrofuran
2-Acetyl-3-hydroxyfuran
2,5-Dimethyl-3-ethylfuran
2-Methoxy-3-isobutyl-5-methylpyrazine

Pyridines

2-Methylpyridine
2-Ethylpyridine
3-Methoxypyridine
2-Ethyltetrahydropyridine
2-Propyltetrahydropyridine

Pyrans

3,4-Dihydropyran
4H-Pyran-2-one
2H-4-Hydroxy-6-methylpyran-2-one

Adapted from Maga, J. A. and Fapojuwo, O. O., *J. Sensory Stud.*, 1, 9, 1986.

Table 22
AROMA DESCRIPTION OF CERTAIN CARBONYLS ISOLATED FROM WOOD SMOKE

Compound	Odor description
2-Cyclopentenone	Grassy, potato-like
3-Methyl-2-cyclopentenone	Somewhat sweet, grassy
2,3-Dimethyl-2-cyclopentenone	Grassy, bitter
2,4-Dimethyl-2-cyclopentenone	Somewhat grassy
2,5-Dimethyl-2-cyclopentenone	Somewhat grassy
3,4-Dimethyl-2-cyclopentenone	Somewhat grassy
3,5-Dimethyl-2-cyclopentenone	Somewhat grassy
3-Ethyl-2-cyclopentenone	Grassy
2-Ethyl-3-methyl-2-cyclopentenone	Mild, medicine-like
2-Ethyl-4-methyl-2-cyclopentenone	Grassy

Adapted from Kim, K., Kurata, T., and Fujimaki, M., *Agric. Biol. Chem.*, 38, 53, 1974.

Table 23
ODOR DESCRIPTIONS OF CERTAIN LACTONES IDENTIFIED IN WOOD SMOKE

Compound	Odor description
γ-Butyrolactone	Somewhat bitter, burnt
2-Methyl-2-butenolide	Sweet, burnt, caramel-like
4-Methyl-2-butenolide	Smoky, burnt
2,3-Dimethyl-2-butenolide	Spicy, vanilla-like
2,4-Dimethyl-2-butenolide	Sweet, burnt
3,4-Dimethyl-2-butenolide	Weak, sour, smoky
2,3,4-Trimethyl-2-butenolide	Weak, burnt, caramel-like
2-Ethyl-4-methyl-2-butenolide	Burnt, woody
4-Ethylidene-2-methyl-2-butenolide	Sweet, caramel-like

Adapted from Kim, K., Kurata, T., and Fujimaki, M., *Agric. Biol. Chem.*, 38, 53, 1974.

Table 24
ACCEPTABILITY OF SMOKE VS. WOOD SOURCE

Good woods:	Beech
	Oak
	Maple
Fair woods:	Pine
	Birch
Poor woods:	Alder
	Lime
	Aspen
	Fir

Adapted from Tilgner, D. J., *Fleischwirtschaft*, 10, 751, 1958.

Table 25
YIELD (mg %) OF TOTAL AND INDIVIDUAL AROMA FRACTIONS AS INFLUENCED BY WOOD SOURCE

	Aroma fraction					
Wood	**Total**	**Phenolic**	**Carbonyl**	**Noncarbonyl**	**Acid**	**Basic**
Oak *(Queretis seriata)*	1860	151	117	99	1140	48
Oak *(Q. acuta)*	1600	225	323	310	820	80
Cherry	1490	101	111	60	660	108
Bamboo	1180	144	86	211	610	40
Pine	1180	166	53	24	640	33
Cedar	890	122	90	73	520	40

Adapted from Fujimaki, M., Kim., K., and Kurata, T., *Agric. Biol. Chem.*, 38, 45, 1974.

Table 26
RELATIVE SMOKE COMPOUND COMPOSITION AS INFLUENCED BY WOOD SOURCE

	% of peak area					
Compound	***Queretis acuta***	***Q. seriata***	**Cherry**	**Bamboo**	**Pine**	**Cedar**
Acetic acid	56	28	9	17	74	10
Propionic acid	13	32	35	50	9	36
Isobutyric acid	3	4	7	3	1	7
n-Butyric acid	1	10	13	15	<0.1	17
Isovaleric acid	<0.1	1	2	1	<0.1	3
cis-Crotonic acid	0.3	<0.1	0.3	<0.1	0.3	2.5
n-Valeric acid	<0.1	1	1	1	—	1
trans-Crotonic acid	0.7	0.7	1.3	0.4	0.2	1.2
Guaiacol	45	40	30	36	26	24
4-Methylguaiacol	12	12	9	7	10	11
4-Ethylguaiacol	2	2	1	2	1	2
o-Cresol	24	23	39	40	40	45
p-Cresol	10	10	13	9	15	16
3,5-Xylenol	4	4	4	3	1	2
3,4-Xylenol	<0.5	<0.5	<0.5	<0.5	<0.5	<0.5
Syringol	4	4	4	3	4	0
4-Methylsyringol	2	1	1	<0.5	1	0
4-Ethylsyringol	0	<0.5	0	0	0	0
Cyclopentanone	3	4	3	11	1	3
2-Methyl-2-cyclopentenone	10	11	13	7	10	11
Furfural	38	38	37	41	49	27
2-Acetylfuran	8	11	9	11	6	9
3-Methyl-2-cyclopentenone	1	1	2	1	3	2
2,3-Dimethyl-2-cyclopentenone	2	2	4	2	2	2
5-Methylfurfural	24	22	22	12	17	31

Adapted from Fujimaki, M., Kim., K., and Kurata, T., *Agric. Biol. Chem.*, 38, 45, 1974.

Table 27
DIFFERENCES BETWEEN THE PHENOLIC COMPOSITION OF HARDWOOD AND SOFTWOOD

	Relative % in	
Compound	**Hardwood**	**Softwood**
Phenol	4	4
o-Cresol	2	2
m-Cresol	2	2
p-Cresol	2	1
Guaiacol	8	10
Pyrocatechol	6	8
4-Methylguaiacol	7	10
4-Methylpyrocatechol	2	4
4-Methylresorcinol	3	4
4-Ethylguaiacol	3	5
Syringol	12	3
4-Vinylguaiacol	3	5
4-Allylguaiacol	2	4
4-Methylsyringol	9	3
4-Ethylsyringol	11	11
4-Vinylsyringol	3	1
4-Allylsyringol	3	1
4-Propenylsyringol	10	2

Adapted from Baltes, W., Wittkowski, R., Sochtig, I., Block, H., and Toth, L., *The Quality of Food and Beverages*, Vol. 2, Charalambous, G. and Inglett, G., Eds., Academic Press, New York, 1981, 1.

Table 28
COMPOSITION OF LIGHT AND HEAVY BEECH AND OAK SMOKES

	Composition in mg/m³ smoke			
Smoke component	**Light beech**	**Light oak**	**Heavy beech**	**Heavy oak**
Ash and soot	7.95	21.77	8.24	23.50
Ether insolubles	216.0	154.16	99.58	754.40
Free bases	45.3	41.96	155.76	48.01
Aldehydes and ketones	47.9	24.75	108.65	70.95
Carboxylic acids	37.28	64.32	203.6	327.5
Phenols	15.91	8.97	93.31	73.27
Neutral organic compounds	298.96	477.60	1757.02	803.64

Adapted from Ziemba, A., *Przem. Spozyw.*, 2, 200, 1957.

Table 29
SMOKE COMPOSITION FROM DIFFERENT WOOD SPECIES

	g/100 g sawdust					
Class/compound	Spruce	Pine	Acacia	Oak	Beech	White beech
Monocarbonyls	3.06	3.10	1.70	2.40	1.77	2.50
Oxy- and dicarbonyls	6.30	6.12	1.86	1.98	3.60	4.95
Formaldehyde	0.21	0.20	0.11	0.14	0.15	0.16
Acetaldehyde	1.15	1.00	0.81	0.63	0.66	0.65
Acetone	0.69	0.54	0.64	0.37	0.74	0.69
Furfural	0.21	0.18	0.24	0.35	0.18	0.26
NaOH reducing substances	7.83	7.40	4.56	4.68	6.94	9.75
Reductants	0.21	0.29	0.13	0.17	0.16	0.32
Dehydroreductants	2.10	1.80	1.56	0.66	1.32	2.16
Total acids	3.07	3.17	2.94	3.56	4.06	6.79
pH	3.8	4.7	4.3	4.0	4.8	4.1
Esters	8.08	11.16	8.48	10.47	6.69	10.35
Phenols	0.29	0.29	0.36	0.16	0.14	0.32

Adapted from Toth, L. and Potthast, K., *Adv. Food Res.*, 29, 87, 1984.

Table 30
INFLUENCE OF WOOD SOURCE ON SMOKED WHITEFISH ACCEPTABILITY

Wood	Sensory hedonic score[a]	Panel comments
Red maple	6.49	Good blend of salty, sweet, smoky
Red oak	6.42	Good; slight sour and burnt
Trembling aspen	6.24	Good; slower flavor impact, sweet
White ash	6.08	Good; slight medicinal
Balsam poplar	6.05	Slower impact; sweet, mild
Birch	5.98	Good; sweet, smoky blend; medicinal
Beech	5.91	Good; flat taste; slightly burnt
White oak	5.82	Fair; slight medicinal, bitter
Hard maple	5.32	Dominant sweet-sour taste
Hickory	5.24	Salt and smoky flavor, poorly blended
Diamond willow	4.70	Strong, fruity, sweet, medicinal, astringent
Burr oak	3.77	Unacceptable; strong medicinal, gasoline-like taste

[a] Maximum = 9.00.

Adapted from Lantz, A. W. and Vaisey, M., *J. Fish Res. Board Can.*, 27, 1201, 1970.

Table 31
INFLUENCE OF WOOD SOURCE ON THE SENSORY PROPERTIES OF SMOKED EEL

Wood	Eel color	Eel odor	Eel flavor
Teak sawdust	Rich red	Acrid	Slightly unpalatable
Teak wood chips	Pale yellow	Acrid	Unpalatable
Sajar sawdust	Reddish yellow	Acrid	Slightly unpalatable
Acacia sawdust	Rich red	Very acrid	Bitter, unpalatable
Mango sawdust	Reddish yellow	Very acrid	Tarry, unpleasant
Coconut husk	Pale yellow	Pleasant	Very palatable
75% coconut husk + 25% teak sawdust	Rich yellow	Pleasant	Palatable
50% coconut husk + 50% teak sawdust	Rich yellow	Pleasant	Palatable
25% coconut husk + 75% teak sawdust	Rich red	Mildly pleasant	Slightly acrid

Adapted from Solanki, K. K., Kandoran, M. K., and Venkataraman, R., *Fish Technol.*, 7, 169, 1970.

Table 32
AROMA INTENSITY RATINGS OF VARIOUS WOOD SMOKE FRACTIONS

Wood source	Carbonyl	Neutral	Basic	Phenolic
Apple	4.02	2.28	5.32	3.40
Chestnut	3.86	1.82	3.44	3.21
Cherry	5.40	4.13	4.61	3.59
Hard maple	4.23	2.75	3.15	3.59
Hickory	4.32	3.40	3.57	4.96
Mesquite	3.15	2.54	2.80	5.83
Red oak	6.03	2.92	4.07	3.17
White oak	5.46	2.63	3.52	3.44

Adapted from Maga, J. A. and Fapojuwo, O. O., *J. Sensory Stud.*, 1, 9, 1986.

Table 33
PYRAZINE CONTENT OF VARIOUS WOOD SMOKE SOURCES

Wood	mg total pyrazines/ 100 g dry wt wood
Redwood	10.33
Chestnut	10.59
Douglas fir (heartwood)	14.08
White oak	17.05
Aspen	17.16
Walnut	17.51
Douglas fir (sapwood)	20.07
Cherry	20.48
Eastern cedar	22.05
Hard maple	22.32
Apple	23.24
Red oak	23.84
Red alder	24.03
Birch	26.28
Lodgepole pine (bark)	29.57
Lodgepole pine (wood)	35.59
Hickory	41.03

Adapted from Maga, J. A. and Chen, Z., *Flavour Fragrance J.*, 1, 37, 1985.

Table 34
INFLUENCE OF SMOLDERING TEMPERATURE ON THE AMOUNTS OF PHENOLS, CARBONYLS, AND ACIDS IN WOOD SMOKE

Smoldering temperature (°C)	Phenols (μg phenol/mℓ)	Carbonyls (mg acetaldehyde/mℓ)	Acids (meq acid/mℓ)
252	88	20.0	0.082
287	91	20.2	0.077
322	153	27.3	0.121
355	197	31.2	0.104
386	165	37.2	0.112

Adapted from Porter, R. W., Bratzler, L. J., and Pearson, A. M., *J. Food Sci.*, 30, 615, 1965.

Table 35
INFLUENCE OF SMOLDERING TEMPERATURE ON THE RECOGNITION THRESHOLD (ppm) AND MOST DESIRABLE CONCENTRATION (ppm) OF VAPOR-PHASE WOOD SMOKE

Smoldering temperature (°C)	Recognition threshold		Most desirable concentration	
	Odor	Taste	Odor	Taste
400	10.4	2.3	20.8	15.6
500	20.8	4.1	41.7	27.1
600	18.7	6.2	37.5	25.2

Adapted from Daun, H., *Lebensm. Wiss. Technol.*, 5, 102, 1972.

Table 36
INFLUENCE OF SMOLDERING TEMPERATURE ON THE PHENOL CONTENT OF BEECHWOOD SMOKE

Smoldering temperature (°C)	Phenol content (mg/100 g)
400	350
450—500	800
500—600	2960
700—800	2810
900—1050	1945

Adapted from Toth, L., *Fleischwirtschaft*, 60, 728, 1980.

Table 37
CHANGE IN VOLATILE COMPOSITION AND CONTENT WITH INCREASING SMOKE GENERATION TEMPERATURE

Generator temperature (°C)	Acids (wt %)	Phenols (wt %)	Carbonyls (wt %)	Total (mg/100 mℓ)
300	41	2.3	57	147
315	40	1.7	59	214
343	32	1.7	66	437
385	27	1.6	72	503
415	24	1.5	74	634

Adapted from Simon, S., Rypinski, A. A., and Tauber, F. W., *Food Technol.*, 20, 1494, 1966.

Table 38
INFLUENCE OF SMOKE GENERATION TEMPERATURE ON THE CONTENT OF CERTAIN PHENOLS

Generation temperature (°C)	mg/100 g			
	Phenol	Guaiacol	Benzocatechol	Syringol
450	20	60	35	85
550	30	85	60	150
650	55	125	135	260
750	20	115	135	215
850	25	100	125	175

Adapted from Toth, L., *Fleischwirtschaft*, 60, 1472, 1980.

Table 39
INFLUENCE OF % OXYGEN IN THE GAS USED FOR SMOLDERING vs. VOLATILE COMPONENT COMPOSITION

	% peak area/g sawdust consumed		
% oxygen in gas	**Furfuryl alcohol**	**Cyclotene**	**Guaiacol**
0	100	55	55
10	75	100	100
20	44	45	55
30	40	43	50
40	38	46	55
50	36	45	48

Adapted from Wasserman, A. E. and Fiddler, W., Proc. Meat Ind. Res. Conf., Chicago, 1969, 163

Table 40
INFLUENCE OF AIR AMOUNT ON THE SENSORY PROPERTIES OF SMOKE VAPOR

	Recognition threshold (ppm)		Most desirable concentration (ppm)	
Air amount	**Odor**	**Taste**	**Odor**	**Taste**
Low	7.4	1.4	15.6	17.7
High	8.3	1.3	16.7	8.3

Adapted from Daun, H., *Lebensm. Wiss. Technol.*, 5, 102, 1972.

Table 41
INFLUENCE OF WOOD MOISTURE CONTENT ON RESULTING CHEMICAL COMPOSITION AND SENSORY PROPERTIES

	Yield/100 g sawdust				
Moisture content (%)	**Condensate (g)**	**Phenols (mg)**	**Acids (mg)**	**Formaldehyde (mg)**	**Sensory quality**
1.8	28.2	236	3293	122	Unsatisfactory
21.5	41.5	136	3288	81	Good
24.5	43.5	100	3003	78	Good
31.2	32.8	33	890	—	Almost satisfactory

Adapted from Gorbatov, V. M., Krylova, N. N., Volovinskaya, V. P., Lyaskouskaya, Y. N., Bazarova, K. I., Khlamova, R. I., and Yakovleva, G. Y., *Food Technol.*, 25, 71, 1971.

Table 42
CHEMICAL COMPARISON OF SAWDUST SMOKE AND FRICTION SMOKE

Property	Sawdust smoke Steam volatile	Sawdust smoke Nonsteam volatile	Friction smoke Steam volatile	Friction smoke Nonsteam volatile
pH	4.00	4.20	3.18	3.00
% total acids	1.28	0.183	4.98	0.209
% total carbonyls	0.627	0.164	4.78	2.79
% total phenols	0.068	0.007	0.217	0.067
% total volatiles	1.35	0.354	9.98	3.07
% moisture	86.7		70.6	

Adapted from Husaini, S. A. and Cooper, G. E., *Food Technol.*, 11, 499, 1957.

Table 43
PHENOLIC DIFFERENCES BETWEEN HOT AND COLD SMOKING

Compound	mg/kg Hot smoke	mg/kg Cold smoke
Cyclotene	Trace	Trace
Guaiacol	Trace	1.90
Maltol	Trace	10.15
Phenol	Trace	2.35
m, *p*-Cresol	Trace	Trace
Eugenol	12.23	34.80
Isoeugenol	Trace	Trace
2,6-Dimethoxy-4-alkylphenol	Trace	Trace
Vanillin	Trace	Trace
Acetovanillone	17.65	Trace
Syringaldehyde	75.25	42.80
Acetosyringone	59.45	Trace

Adapted from Potthast, K., *Acta Aliment. Pol.*, 3, 189, 1977.

Table 44
EFFECT OF SMOKE AGE AS COMPARED TO A 2-MONTH-OLD CONTROL

Age	Score (7 point hedonic)
2 month	4.0
1 month	4.1
96 hr	3.6
48 hr	3.6
5 hr	2.8

Adapted from Wasserman, A. E. and Fiddler, W., Proc. Meat Ind. Res. Conf., Chicago, 1969, 163.

Table 45
INFLUENCE OF STORAGE ON METHYL ACETATE AND FORMATE LEVELS

	Concentration (molarity)	
Storage time (hr)	**Methyl acetate**	**Methyl formate**
0	0	0.001
100	0.010	0.009
200	0.027	0.010
300	0.036	0.011
400	0.037	0.009
500	0.038	0.009
600	0.039	0.010
700	0.041	0.010

Adapted from Doerr, R. C., Wasserman, A. E., and Fiddler, W., *J. Agric. Food Chem.*, 14, 662, 1966.

REFERENCES

1. **Fiddler, W., Wasserman, A. E., and Doerr, R. C.,** A smoke flavor fraction of a liquid smoke solution, *J. Agric. Food Chem.*, 18, 934, 1970.
2. **Knowles, M. E., Gilbert, J., and McWeeny, D. J.,** Phenols in smoked cured meats. Phenolic composition of commercial liquid smoke preparations and derived bacon, *J. Sci. Food Agric.*, 26, 189, 1975.
3. **Kornreich, M. R. and Issenberg, P.,** Determination of phenolic wood smoke components as trimethylsillyl esters, *J. Agric. Food Chem.*, 20, 1109, 1972.
4. **Kim, K., Kurata, T., and Fujimaki, M.,** Identification of flavor constitutents in carbonyl, non-carbonyl, neutral and basic fractions of aqueous smoke condensates, *Agric. Biol. Chem.*, 38, 53, 1974.
5. **Lustre, A. O. and Issenberg, P.,** Phenolic components of smoked meat products, *J. Agric. Food Chem.*, 18, 1056, 1970.
6. **Fiddler, W., Doerr, R. C., and Wasserman, A. E.,** Composition of an ether-soluble fraction of a liquid smoke solution, *J. Agric. Food Chem.*, 18, 310, 1970.
7. **Radecki, A., Lamparczyk, H., Grzybowski, J., Halkiewicz, J., Bednarek, P., Boltrukiewicz, J., and Rapicki, W.,** Analysis of a commercial smoke-curing agent. III. Separation and identification of components of low-boiling phenolic fractions, *Bromatol. Chem. Toksykol.*, 9, 327, 1976.
8. **Radecki, A., Grzybowski, J., Halkiewicz, J., Lamparcyzk, H., Bednarek, P., Boltrukiewicz, J., and Rapicki, W.,** Analysis of a commercial smoke preparation. IV. Separation and identification of some components significantly influencing the organoleptic and preserving properties of the preparation, *Bromatol. Chem. Toksykol.*, 9, 461, 1976.
9. **Radecki, A., Bednarek, P., Boltrukiewicz, J., Cacha, R., Rapicki, W., Grzybowski, J., Halkiewicz, J., and Lamparczyk, H.,** Analysis of a commercial smoke-curing agent. I. Separation and identification of high-boiling phenolic fractions, *Bromatol. Chem. Toksykol.*, 8, 179, 1975.
10. **Radecki, A., Halkiewicz, J., Lamparczyk, H., Grzybowski, J., Bednarek, P., Boltrukiewicz, J., Cacha, R., and Rapicki, W.,** Analysis of a commercial smoke-curing agent. II. Separation and identification of some constitutents of the low-boiling phenolic fraction, *Bromatol. Chem. Toksykol.*, 8, 185, 1975.
11. **Radecki, A. and Grzybowski, J.,** Method for determination of phenolic compounds in smoked products, *Bromatol. Chem. Toksykol.*, 14, 129, 1981.
12. **Ishiwata, H., Watanabe, H., Watanabe, M., Hayashi, T., Hara, Y., Kato, S., and Tanimura, A.,** Chemical compositions of smoke flavour, *Bull. Natl. Inst. Hygienic Sci.*, 94, 112, 1976.
13. **Hoff, J. E. and Kapsalopoulou, A. J.,** Low-boiling components of wood smoke. Identification by means of syringe reactions, *J. Gas Chromatogr.*, 2, 296, 1964.
14. **Doerr, R. C., Wasserman, A. E., and Fiddler, W.,** Composition of hickory sawdust smoke. Low-boiling constituents, *J. Agric. Food Chem.*, 14, 662, 1966.

15. **Love, S. and Bratzler, L. J.**, Tentative identification of carbonyl compounds in wood smoke by gas chromatography, *J. Food Sci.*, 31, 218, 1966.
16. **Hamid, H. A. and Saffle, R. L.**, Isolation and identification of the volatile fatty acids present in hickory sawdust smoke, *J. Food Sci.*, 30, 697, 1965.
17. **Radecki, A., Grzybowski, J., Halkiewicz, J., and Lamparczyk, H.**, Isolation and identification of some components of the lower-boiling fraction of a commercial smoke flavoring, *Acta Aliment. Pol.*, 3, 203, 1977.
18. **Porter, R. W., Bratzler, L. J., and Pearson, A. M.**, Fractionation and study of compounds in wood smoke, *J. Food Sci.*, 30, 615, 1965.
19. **Luten, J. B., Ritskes, J. M., and Weseman, J. M.**, Determination of phenol, guaiacol and 4-methylguaiacol in wood smoke and smoked fish products by gas-liquid chromatography, *Z. Lebensm. Unters. Forsch.*, 168, 289, 1979.
20. **Hurza, D. E., van Praag, M., and Heinsohn, H.**, Isolation and identification of the components of tar of hickory wood smoke, *J. Agric. Food Chem.*, 22, 123, 1974.
21. **Lustre, A. O. and Issenberg, P.**, Volatile components of hardwood sawdust smoke, *J. Agric. Food Chem.*, 17, 1387, 1969.
22. **Husaini, S. A. and Cooper, G. E.**, Fractionation of wood smoke and the comparison of chemical composition of sawdust and friction smokes, *Food Technol.*, 11, 499, 1957.
23. **van Chuyen, N. and Kato, H.**, Recent studies on smoke flavor. II, *Nippon Shokuhin Kogyo Gakkai Shi*, 30, 722, 1983.
24. **Borys, A., Flossowska, B., Obiedzinski, M., and Olkiewicz, M.**, Influence of combustion conditions on the composition of carbonylic, carboxylic, polynucelar hydrocarbons and phenolic compounds present in wood smoke, *Acta Aliment. Pol.*, 3, 335, 1977.
25. **Fiddler, W., Doerr, R. C., Wasserman, A. E., and Salay, J. M.**, Composition of hickory sawdust smoke. Furans and phenols, *J. Agric. Food Chem.*, 14, 659, 1966.
26. **Potthast, K.**, Determination of phenols in smoked meat products, *Acta Ailment. Pol.*, 3, 189, 1977.
27. **Toth, L.**, Separation and analysis of phenol fractions from smokehouse smoke. II. Separation, purification and analysis of phenol fractions, *Fleischwirtschaft*, 60, 728, 1980.
28. **Jodai, S.**, Studies of smoke perfume. I. Organic solvent extracts of pyroligneous acid, *J. Food Sci. Technol.*, 15, 335, 1968.
29. **Fujimaki, M., Kim, K., and Kurata, T.**, Analysis and comparison of flavor constituents in aqueous smoke condensates from various woods, *Agric. Biol. Chem.*, 38, 45, 1974.
30. **Borys, A.**, Composition of carbonyl fraction in smoke from selected types of wood, *Rocz. Inst. Przem. Miesnego Tluszezowego*, 15, 157, 1978.
31. **Kasahara, K. and Nishibori, K.**, Volatile components of smoked salmon. I, *Bull. Jpn. Soc. Sci. Fish.*, 45, 1543, 1979.
32. **Kasahara, K. and Nishibori, K.**, Volatile components of smoked salmon. II. The relationship between flavors of smoked salmon and smoke tar, *Bull. Jpn. Soc. Sci. Fish.*, 45, 1547, 1979.
33. **Kasahara, K. and Nishibori, K.**, Volatile components of smoked cuttlefish and octopus, *Bull. Jpn. Soc. Sci. Fish*, 49, 913, 1983.
34. **Kasahara, K. and Nishibori, K.**, Volatile components of smoked herring, *Bull. Jpn. Soc. Sci. Fish.*, 47, 783, 1981.
35. **Kasahara, K. and Nishibori, K.**, Volatile components of smoked sardine, *Bull. Jpn. Soc. Sci. Fish.*, 47, 113, 1981.
36. **Kasahara, K. and Nishibori, K.**, Volatile carbonyls and acids of smoked sardines, *Bull. Jpn. Soc. Sci. Fish.*, 48, 691, 1982.
37. **Borys, A.**, Chemical composition of the Bieszczadzki refined liquid smoke preparation, *Rocz. Inst. Przem. Miesnego Tluszczowego*, 16, 95, 1979.
38. **Toth, L.**, Influence of smokehouse technology on phenols in the smoke, *Fleischwirtschaft*, 60, 1472, 1980.
39. **Wasserman, A. E. and Fiddler, W.**, Natural smoke: composition and properties, Proc. Meat Ind. Res. Conf., Chicago, 1969, 163.
40. **Baltes, W. and Sochtig, I.**, Detection by glass capillary gas chromatography of addition of smoke flavor condensate to sausages, *Z. Lebensm. Unters. Forsch.*, 169, 17, 1979.
41. **Fiddler, W., Parker, W. E., Wasserman, A. E., and Doerr, R. C.**, Thermal decomposition of ferulic acid, *J. Agric. Food Chem.*, 15, 757, 1967.
42. **Olsen, C. Z.**, Chemical composition and application of smoke flavor, *Proc. Eur. Meet. Meat Res. Workers*, No. 22, F7:1, 1976.
43. **Baltes, W. and Sochtig, I.**, Low molecular weight ingredients of smoke flavor preparations, *Z. Lebensm. Unters. Forsch.*, 169, 9, 1979.
44. **Borys, A., Obiedzinski, M., and Olkiewicz, M.**, Identification of phenolic compounds in a polish smoke flavor extract, *Proc. Eur. Meet. Meat. Res. Workers*, No. 22, F8:1, 1976.

45. **Wittkowski, R., Toth, L., and Baltes, W.**, Separation and analysis of phenol fractions from smokehouse smoke. III. Separation and identification of mono- and di-hydroxycompounds, *Z. Lebensm. Unters. Forsch.*, 173, 445, 1981.
46. **Baltes, W., Block, H., and Mevissen, L.**, Studies of high molecular weight tars from smoke aroma condensates, *Z. Lebensm. Unters. Forsch.*, 175, 130, 1982.
47. **Lapshin, I. I. and Rodina, T. G.**, The role of phenols in the MINCH liquid smoke separation for treatment of fish, *Rybn. Khoz.*, No. 2, 75, 1974.
48. **Goos, A. W.**, The thermal decomposition of wood, *Wood Chemistry*, Vol. 2, 2nd ed., Reinhold, New York, 1952.
49. **McWeeny, D. J.**, Reactions of some food additives during storage, *Food Chem.*, 9, 89, 1982.
50. **Bratzler, L. J., Spooner, M. E., Weatherspoon, J. B., and Maxey, J. A.**, Smoke flavor as related to phenol, carbonyl and acid content of bologna, *J. Food Sci.*, 34, 146, 1969.
51. **Issenberg, P., Kornreich, M. R., and Lustre, A. O.**, Recovery of phenolic wood smoke components from smoked foods and model systems, *J. Food Sci.*, 36, 107, 1971.
52. **Wasserman, A. E.**, Organoleptic evaluation of three phenols present in wood smoke, *J. Food Sci.*, 31, 1005, 1966.
53. **Daun, H.**, Sensory properties of phenolic compounds isolated from curing smoke as influenced by its generation parameters, *Lebensm. Wiss. Technol.*, 5, 102, 1972.
54. **Maga, J. A.**, Pyrazines in foods. An update, *CRC Crit. Rev. Food Sci. Nutr.*, 16, 1, 1982.
55. **Tilgner, D. J.**, Analysis and use of smoke from various kinds of wood, *Fleischwirtschaft*, 10, 751, 1958.
56. **Khvan, E. A., Shaposhnikov, Y. K., Radakova, T. N., Alsufev, V. A., and Kondakova, L. V.**, Study of the chemical composition of smoke evolved by different types of wood, *Rybn. Khoz.*, No. 5, 62, 1975.
57. **Lee, J. M. and Song, K. W.**, Effects of conventional smoking and liquid smoke treatment on the quality of pork sausages, *Korean J. An. Sci.*, 17, 511, 1975.
58. **Sidaway, E. P.**, A comparison of some British Columbia softwoods for smoking fish, *Fish. Res. Board Can. Prog. Rep. Pac. Coast Stn.*, 59, 12, 1944.
59. **Lantz, A. W. and Vaisey, M.**, Flavor effects of different woods on whitefish smoked in a kiln with controlled temperature, humidity, and air velocity, *J. Fish. Res. Board Can.*, 27, 1201, 1970.
60. **Solanki, K. K., Kandoran, M. K., and Venkataraman, R.**, Studies on smoking of eel fillets, *Fish Technol.*, 7, 169, 1970.
61. **Simon, S., Rypinski, A. A., and Tauber, F. W.**, Water-filled cellulose casings as model absorbents for wood smoke, *Food Technol.*, 20, 1494, 1966.
62. **Klossowska, B. M.**, Wood destruction rate versus combustion conditions in the process of generating wood smoke — a mathematical approach, *Acta Ailment. Pol.*, 7, 283, 1981.
63. **Gorbatov, V. M., Krylova, N. N., Volovinskaya, V. P., Lyaskovskaya, Y. N., Bazarova, K. I., Khlamova, R. I., and Yakovleva, G. Y.**, Liquid smokes for use in cure meats, *Food Technol.*, 25, 71, 1971.
64. **Baltes, W., Wittkowski, R., Sochtig, I., Block, H., and Toth, L.**, Ingredients of smoke and smoke flavor preparations, in *The Quality of Foods and Beverages*, Vol. 2, Charalambous, G. and Inglett, G., Eds., Academic Press, New York, 1981, 1.
65. **Maga, J. A. and Sizer, C. E.**, Pyrazines in foods. A review, *J. Agric. Food Chem.*, 21, 22, 1973.
66. **Maga, J. A. and Sizer, C. E.**, Pyrazines in foods, *CRC Crit. Rev. Food Technol.*, 4, 39, 1973.
67. **Maga, J. A.**, Pyrazines in foods. An update, *CRC Crit. Rev. Food Sci. Nutr.*, 16, 1, 1982.
68. **Maga, J. A.**, Flavor contribution of wood in alcoholic beverages, in *Progress in Flavour Research*, Adda, J., Ed., Elsevier, Amsterdam, 1985, 409.
69. **Maga, J. A. and Fapojuwo, O. O.**, Aroma intensities of various wood smoke fractions, *J. Sensory Stud.*, 1, 9, 1986.
70. **Maga, J. A. and Chen, Z.**, Pyrazine composition of wood smoke as influenced by wood source and smoke generation variables, *Flavour Fragrance J.*, 1, 37, 1985.
71. **Maga, J. A.**, Lactones in foods, *CRC Crit. Rev. Food Sci. Nutr.*, 8, 1, 1976.
72. **Ziemba, A.**, Generation of industrial curing smoke and its chemical composition, *Przem. Spozyw.*, 2, 200, 1957.
73. **Toth, L. and Potthast, K.**, Chemical aspects of the smoking of meat and meat products, *Adv. Food Res.*, 29, 87, 1984.

Chapter 5

TEXTURAL CHANGES IN SMOKED FOODS

I. INTRODUCTION

One would assume that the smoking of foods could have a significant influence on their textural properties but one has difficulty in designing a valid experiment to investigate the problem. As pointed out by Sokolov et al.,[1] textural properties in products such as sausages can be influenced by a multitude of factors including alteration of original muscle structure due to grinding; emulsion formation due to protein activity via hydrogen-, sulfide-, and ionic bonding; action of muscle and microbial proteases; influence of temperature and moisture content during smoking; and the interaction among smoke compounds and food components. In addition, pH can interact with numerous of the above factors to indirectly influence product texture.

II. PROTEIN INTERACTIONS

It is a well-accepted fact that heat will result in protein denaturation or loss of solubility, and since protein is the major reactive food component contributing to the textural properties of muscle-based foods, it is not surprising that most smoke treatments which also involve some degree of heat processing alter textural properties.[2]

Data presented by Randall and Bratzler,[3] as summarized in Table 1, clearly demonstrate that smoking of muscle has an additional influence over that of heat itself in altering protein composition, and thus textural properties. As can be seen, smoking significantly lowered both myofibrillar and sarcoplasmic protein nitrogens while increasing stromal protein nitrogen. Similar results were obtained with other meat protein sources.[4]

These protein changes apparently result in a cross-linking of surface proteins resulting in a firm and stable outer crust which texturally is harder or tougher than the softer inside portion of the product. The formation of this crust also hinders the penetration of smoke constituents toward the center of the product, thus resulting in uneven distribution of color and flavor.

III. SMOKE SOURCE

Several studies have clearly demonstrated that smoke source and/or application technique can also influence the textural properties of smoked foods. For example, a portion of the data generated by Breclaw and Dawson[5] is summarized in Table 2, and as can be seen, type of smoke used and storage time did influence product tenderness. These changes would probably correlate with changes in protein solubility. They also performed similar experiments with darker poultry meat, which was found to be consistently tougher than white meat, again apparently due to differences in protein composition between light and dark meat.

However, Janky et al.,[6,7] in working with Cornish game hen light and dark meat, reported that dark meat was more tender than light meat. However, protein compositional data were not provided. They also clearly demonstrated that smoking resulted in a more tender product, regardless of color, than an oven-roasted control.

Sink and Hsu[8] as can be seen in Table 3, also demonstrated that smoke source can significantly influence the textural properties of foods. Liquid smoke that was mixed with frankfurter emulsion produced a significantly softer product than the other liquid or solid

Table 1
INFLUENCE OF SMOKING ON PORK MUSCLE PROTEIN FRACTION % NITROGEN COMPOSITION

Treatment	% myofibrillar protein nitrogen	% sarcoplasmic protein nitrogen	% stromal protein nitrogen
Control (unheated)	50.3	17.7	19.8
Control (heated, no smoke)	61.0	9.6	18.0
Heated (smoke, aerosol)	41.3	5.4	41.2

Adapted from Randall, C. J. and Bratzler, L. J., *J. Food Sci.*, 35, 248, 1970.

Table 2
INFLUENCE OF SMOKE PROCESS AND STORAGE (1°C) ON WHITE POULTRY MEAT TENDERNESS

	Shear value peak heights (mm)		
Treatment	**1 Day**	**7 Days**	**14 Days**
Control (no smoke)	64.1	46.6	57.5
Liquid smoke (brine)	68.3	64.7	63.5
Liquid smoke (dip)	63.0	57.7	68.6
Solid smoke (aerosol)	56.4	61.4	71.6

Adapted from Breclaw, E. W. and Dawson, L. E., *J. Food Sci.*, 35, 379, 1970.

Table 3
INFLUENCE OF SMOKE PROCESS ON FRANKFURTER TENDERNESS

Treatment	Shear score (kg)
Liquid smoke (mix)	1.67[a]
Solid smoke (aerosol)	1.80
Liquid smoke (aerosol)	1.91[b]
Liquid smoke (dip)	1.99[b]

[a] Significantly more tender.
[b] Significantly more tough.

Adapted from Sink, J. D. and Hsu, L. A., *J. Food Sci.*, 42, 1489, 1977.

smoke applications. No statistical difference was found between liquid smoke applied as an aerosol vs. as a dip.

IV. pH AND WATER-HOLDING CAPACITY (WHC)

A. pH

Reports have appeared showing that smoke source will also influence final product pH,[3,9]

Table 4
INFLUENCE OF SMOKE SOURCE AND FOOD PRODUCT ON pH

Treatment	Product	pH	Ref.
Solid smoke (aerosol)	Pork	4.95	3
	Frankfurter	6.08	9
Control (no smoke)	Pork	5.48	3
Liquid smoke (aerosol)	Frankfurter	5.85	9
Liquid smoke (dip)	Frankfurter	5.61	9
Liquid smoke (mix)	Frankfurter	6.02	9

which as mentioned in the introduction to this chapter, can in turn influence the textural properties of a product primarily due to its ionic influences on salts in the product, which in turn can influence the WHC of the product. However, care must be taken in the interpretation of smoke source vs. pH because the type of food smoked must also be considered. As can be seen in Table 4, with pork,[3] solid smoke produced the lowest pH, while in a frankfurter formulation[9] it produced the highest pH. In any event, it is a factor to consider.

B. WHC

Radetic et al.[10,11] have attempted to correlate smoke composition relative to carbonyl and phenol content to pH and WHC in smoked beef. Essentially they found that WHC was inversely related to the concentration of carbonyls and phenols in smoked meat, and that pH was not significantly associated with WHC. Thus, areas of meat at or near the surface had high levels of carbonyls and phenols as compared to inside locations, but the outside portion had low WHC probably due to the fact that smoke compounds had interacted with protein thus not permitting protein to hold large amounts of water. In turn, this influenced the textural properties with the surface material being rather dry and tough as compared to the interior portion.

REFERENCES

1. **Sokolov, A. A., Tschechowskaja, W. T., and Chekhovskaya, V. T.,** Structure formation in smoked and non-smoked raw sausage, *Fleisch,* 30(1), 15, 1976.
2. **Kihara, S.,** Changes in chicken plasma protein during preservation and processing. III. Absorption of smoke ingredients and the changes in plasma protein and free amino acids during smoking, *Tokyo Nogyo Daigaku Nogakubu Shuho,* 8, 146, 1962.
3. **Randall, C. J. and Bratzler, L. J.,** Changes in various protein properties of pork muscle during the smoking process, *J. Food Sci.,* 35, 248, 1970.
4. **Randall, C. J. and Bratzler, L. J.,** Effect of smoke process solubility and electrophoretic behavior of meat proteins *J. Food Sci.,* 35, 245, 1970.
5. **Breclaw, E. W. and Dawson, L. E.,** Smoke-flavored chicken rolls, *J. Food Sci.,* 35, 379, 1970.
6. **Janky, D. M., Oblinger, J. L., and Koburger, J. A.,** Sensory and microbiological evaluation of smoked cornish game hens, *Poultry Sci.,* 54, 1942, 1975.
7. **Janky, D. M., Koburger, J. A., and Oblinger, J. L.,** Effect of salt brining and cooking procedure on tenderness and microbiology of smoked cornish game hens, *Poultry Sci.,* 55, 761, 1976.
8. **Sink, J. D. and Hsu, L. A.,** Chemical effects of smoke-processing on frankfurter quality and palatability characteristics, *Meat Sci.,* 3, 247, 1979.
9. **Sink, J. D. and Hsu, L. A.,** Chemical effects of smoke-processing on frankfurter manufacture and storage characteristics, *J. Food Sci.,* 42, 1489, 1977.
10. **Radetic, P. M., Suvakov, M. D., and Panin, J. T.,** Influence of carbonylic and phenolic compounds of smoke on pH and water holding capacity of beef, *Proc. Eur. Meet. Meat Res. Workers,* 1(28), 185, 1982.
11. **Radetic, P. M., Suvakov, M. D., and Panin, J. T.,** Effect of carbonyl and phenolic components of smoke on pH and water binding capacity of smoked beef, *Tehnol. Mesa,* 23, 194, 1982.

Chapter 6

THE NUTRITIONAL QUALITY OF SMOKED FOODS

I. INTRODUCTION

Any type of processing usually has an impact on certain aspects of the nutritional components of food. However, we should remember that these changes can be either positive or negative, depending upon the specific parameter being evaluated.

The actual number of studies that have specifically been concerned with the role of smoke on the resulting nutritional quality in food systems has been relatively few, which is somewhat surprising, since nutrient composition and functionality are becoming major concerns among consumers. Overall, the data in this area are somewhat lacking, thereby making this a relatively important aspect of any planned future research.

II. AMINO ACID COMPOSITION

A. Available Lysine

From a nutritional standpoint, most smoked foods can be considered to be high in protein and thus total protein content is not usually a primary matter of concern. However, what can be of concern is the stability of essential amino acids to the smoking process.

Usually the amino acid of major concern is lysine, since for quite a few foods it is the essential amino acid present in the lowest amount. Also, lysine is a very reactive compound and as such can enter into many chemical reactions common to many types of food processing. All studies that have specifically looked at lysine levels during the smoke process have reported losses in available lysine.[1-4,20-22]

As would be expected, factors such as smoking time, smoking temperature, storage time, and water activity during storage can all influence the amount of available lysine lost. For example, Dvorak and Vognarova[1] concluded that short smoking time did not influence lysine loss but valid data as to the actual relationship of smoking time vs. available lysine loss are not readily available. However, a loss approaching 44% was reported for salami that was cold smoked for 2 days and stored at 20°C for 1 year.[1] However, it is not known how much of this loss was due to smoking and how much was due to the storage effect.

Chen and Issenberg[3] also reported losses of up to 44% in available lysine for uncured beef that was smoked at 65°C for 10 hr, whereas samples that were processed under similar conditions without the application of smoke only lost 15% available lysine, thereby indicating that smoke does lower lysine availability as compared to nonsmoke processing. By exposing beef homogenates to various smoke fractions, Chen and Issenberg[3] demonstrated the phenolic and neutral fractions of smoke apparently contained the compounds that were primarily responsible for loss of available lysine. These data are summarized in Table 1. As pointed out by the authors, these various smoke fractions were far from pure, and thus, the same compounds could have been present in more than one fraction.

The authors did attempt to evaluate the influence of individual compounds known to be present in the various smoke fractions relative to their ability to react with lysine in a bovine serum albumin model system. Specific compounds evaluated included formaldehyde, glyoxal, pyruvaldehyde, coniferaldehyde, sinapaldehyde, furfuryl alcohol, phenol, eugenol, and syringol. Their data supported the contention of Dvorak and Vognarova[1] that formaldehyde is one of the most reactive smoke by-products relative to action on available lysine. Also, the phenol-type compounds evaluated did not apparently react with lysine. This in turn supports the earlier conclusions of Krylova et al.[5] and Kako[6] that smoke-associated carbonyl com-

Table 1
INFLUENCE OF SMOKE FRACTIONS ON AVAILABLE LYSINE LOSS IN BEEF

Treatment	Available lysine loss (%)
Beef and acidic smoke fraction	14
Beef and phenolic smoke fraction	38
Beef and neutral smoke fraction	45

Adapted from Chen, L. B. and Issenberg, D., *J. Agric. Food Chem.*, 20, 1113, 1972.

pounds react with protein amino groups whereas phenols primarily react with protein sulfhydryl groups.

From a practical standpoint, significant losses of sulfur-containing amino acids apparently do occur as a result of smoking, at least in a survey for various smoked sausages, since all products evaluated were found to be low in lysine.[8]

One report has also appeared in which it was suggested that dipping in liquid smoke resulted in less available lysine loss than found with the use of tradiational smoking.[9] The authors postulated that this resulted from less protein aggregation which is thought to occur with the former form of smoke application. This claim deserves additional research.

B. Methionine

Using a frankfurter model system, Strange et al.[7] demonstrated that the smoking process itself did not significantly influence methionine level, but the use of high peroxide fat in the formulation through subsequent oxidation converts methionine to methionine sulfoxide in the raw product. However, the subsequent cooking/smoking process reduces this compound back to the original methionine level.

In contrast to the above studies, a recent study by Seuss[23] evaluated the total amino acid content of thin frankfurters that were processed from smoke derived from smoldering wood, friction, or smoke condensate. It was reported that only slight changes in amino acid content were observed, but it must be remembered that the product, because of its size, was only exposed to smoke for relatively short periods of time. It should also be remembered that amino acid losses, when they are reported to occur, are more prevalent at the surface of the product with little or no losses apparently occurring in the interior of the product.

III. PROTEIN DIGESTIBILITY

Aside from individual amino acid composition as discussed above, another major nutritional issue is how digestible is the protein associated with smoked foods. Most researchers are in general agreement that the smoking of foods apparently improves protein digestibility.[10-14] However, the actual mechanism for this observation is not clearly understood. Some believe that especially with storage, smoke promotes hydrolysis, thereby improving protein digestibility,[10,11] while others more specifically postulate that certain unidentified smoke components serve as enzyme activators, which in turn promote protein hydrolysis.[14]

Several groups have reported that the use of liquid smoke gives better protein digestibility results than traditional smoke.[12,14,19] One would assume that this is due to the fact that, in general, liquid smoke can come into more direct and complete contact with muscle than traditional smoke and thus results in more protein hydrolysis.

It should be noted that the above studies were conducted in vitro. In vivo studies have been reported,[4,15] but data of these types have to be interpreted rather carefully since they indicate that treating protein with smoke lowers the overall nutritional value of the product in question. It should be remembered that the formation of antinutritional factors in smoke are not effectively considered in most in vivo studies, and thus, factors such as temperature and duration of smoking can influence protein utilization in a negative manner.

IV. PROTEIN ALTERATION

One should also remember that in most cases the smoking of a protein-based food is also accompanied by some degree of heat treatment, and thus, it may be difficult to qualitate and quantitate compositional changes associated with proteins that are specifically due to the influence of smoke and its components. Actually, this represents a fascinating area of research that deserves further attention so that the specific role of smoke on protein functionality can be better understood.

In working with a fish protein source, Abdurakhmanov and Aleiv[16] reported that during smoking, the quantity of muscle albumin did not change, while the amount of low molecular weight proteins increased, and the myofibrillar protein fraction decreased. In addition, the amount of nonprotein nitrogen was found to increase. Thus, from this study it would appear that an interesting study would be to fractionate wood smoke to determine classes and/or types of protein-wood smoke components that contribute most to protein alteration and/or denaturation. These data in turn could be applied to perhaps explain some of the textural properties associated with smoked meats.

V. VITAMINS

Other nutrients that have received some attention relative to the influence of smoke treatment include certain vitamins, especially B vitamins. In the case of various fish-based cured, smoked, and sterilized foods, Wituszynska[17] reported that the riboflavin, nicotinic acid, pantothenic acid, and vitamin B_6 levels decreased aproximately 50% in going from the raw to smoked stages, with subsequent sterilization resulting in another 10% loss. In the opinion of the current author, this represents significant losses and again model systems involving the reactivity of various vitamins with fractionated smoke should be performed in an effort to better understand the actual chemistry involved. This could also lead to ways to minimize such losses, since it is generally conceded that normal heat processing does not dramatically lower B-vitamin content of most foods.

Apparently, B-vitamin losses can vary dependent upon food type since Krylova et al.[18] demonstrated that with pork only minimal losses were observed. A portion of their data are summarized in Table 2, and as can be seen, traditional smoking resulted in losses approaching 15% over a no-smoke, roasted control, while the utilization of liquid smoke resulted in only minimal loses. The differences seen between traditional and liquid smoke were not postulated upon by the authors, but perhaps differences in moisture content between the two experimental treatments could be a possible explanation.

Daun[24] has reported that smoked flesh foods can lose 2 to 25% of their thiamine while losses in niacin and riboflavin are negligible. Probably these losses are due to heat as much as the actual smoke.

Table 2
INFLUENCE OF SMOKE TYPE AND TEMPERATURE ON RIBOFLAVIN AND THIAMINE CONTENT IN PORK

Preparation method	mg% riboflavin	mg% thiamine
Roasted, no smoke	0.080	0.74
Roasted, smoked	0.071	0.65
Roasted, liquid smoke	0.076	0.72

Adapted from Kyrlova, N. N., Karmyshova, L. G., and Kolesnikova, V. T., *Myasn. Ind.*, 9, 36, 1980.

REFERENCES

1. **Dvorak, Z. and Vognarova, I.**, Available lysine in meat and meat products, *J. Sci. Food Agric.*, 16, 305, 1965.
2. **Inagami, K. and Horii, M.**, Changes in available lysine in food protein by heating and smoking, *Kyushu Daigaku Nogak ubu Gakugel Zasshi*, 22, 191, 1966.
3. **Chen, L. B. and Issenberg, P.**, Interactions of some wood smoke components with ε-amino groups in protein, *J. Agric. Food Chem.*, 20, 1113, 1972.
4. **di Cesare, L. F., Sensei, E., and Moioii, F.**, Effects of smoking on the available lysine content and in-vitro digestibility of fillets of *Sardine pilchardus*, *Technol. Aliment*, 3(7), 25, 1980.
5. **Krylova, N. N., Bazarova, L. I., and Kuzeniaova, V. V.**, Interaction of smoke components with meat constituents, presented at *8th Eur. Meet Meat Res. Workers*, Moscow, Aug. 20—27, 1970.
6. **Kako, Y.**, Studies on muscle proteins, II. Changes of beef-, pork-, and chicken-protein during meat product manufacturing processes, *Mem. Fac. Agric. Kagoshima*, 6, 175, 1968.
7. **Strange, E. D., Benedict, R. C., and Miller, A. J.**, Effect of processing variables on the methionine content of frankfurters, *J. Food Sci.*, 45, 632, 1980.
8. **Dubinskaya, A. P., Kulikova, V. V., and German, N. A.**, Nutritional value of cooked sausages, *Izv. Vyssh. Uchebn. Zaved. Pishch. Tekhnol.*, 5, 163, 1978.
9. **Sokolov, A. A., Chekhouskaya, V. T., and Rodin, O. S.**, Food value of freshly cured and freshly smoked sausages, *Izv. Vyssh. Uchebn. Zaved. Pishch. Tekhnol.*, 2, 67, 1973.
10. **Nicora, L. M., Lambert, R., and Wiaux, A.**, Chemical alterations in smoked cured meats during storage, *Ind. Aliment. Agric.*, 83, 1623, 1966.
11. **Sokolov, A. A. and Korobkova, N. N.**, Changes in the nutritional value of raw sausage during curing, *Izv. Vyssh. Uchebn. Zaved. Pishch. Tekhnol.*, 5, 41, 1976.
12. **Khlamova, R. I. and Soloveva, L. I.**, Effect of smoking methods on degradation of pork proteins by proteolytic enzymes, *Myasn. Ind.*, 6, 41, 1978.
13. **Kolesnikova, V. T.**, The digestibility (in vitro) of the muscle proteins of pork products as affected by processing, *Proc. Eur. Meet Meat Res. Workers*, 25, 369, 1979.
14. **Krylova, N. N., Karmyshova, L. G., and Kolesnikova, V. T.**, Effect of liquid smoke preparations on in vitro digestibility of pork loin muscle tissue proteins, *Proc. Eur. Meet. Meat Res. Workers*, 28, 11, 592, 1982.
15. **Hoffman, A., Barranco, A., Francis, B. J., and Disney, J. G.**, The effect of processing and storage upon the nutritive value of smoked fish from Africa, *Trop. Sci.*, 19, 41, 1977.
16. **Abdurakhmanov, Y. A. and Aliev, V. A.**, Effect of processing anchovy kilka on fraction composition of its proteins, *Rybn. Khoz.*, 3, 69, 1976.
17. **Wituszynska, B.**, Effect of various technological processes on the content of B group vitamins in fish preserves, *Bromatol. Chem. Toksykol.*, 9, 467, 1976.
18. **Krylova, N. N., Karmyshova, L. G., and Kolesnikova, V. T.**, Dependence of vitamin content of brisket on the method of preparation, *Myasn. Ind.*, 9, 36, 1980.
19. **Krylova, N. N., Karmyshova, L. G., Balabukh, A. A., Kolesnikova, V. T., Khlamova, R. I., Soloveva, L. I., and Petakova, A. N.**, Quality and nutritive value of pig meat products smoked or treated with a liquid smoke preparation, *Proc. Eur. Meet. Meat Res. Workers*, 24, 9:1, 1978.

20. **Clifford, M. N., Tang, S. L., and Eyo, A. A.,** The development of analytical methods for investigating chemical changes during fish smoking, in *Advances in Fish Science and Technology*, Unipub, New York, 1980, 286.
21. **Petrichenko, L. K. and Dolzhenko, A. A.,** Changes in the amino acid composition of muscle proteins of cyplidnids during smoking, *Rybn. Khoz.*, 6, 71, 1981.
22. **Cattaneo, P., Balzaretti, C., Basilico, C., Beretta, G., and Cantoni, C.,** Production of nutritive value of smoked *Salmo gardneri, Ind. Aliment.*, 531, 1983.
23. **Seuss, I.,** Influence of smoking on nutritionally important components of meat products. II. Protein digestibility and amino acid content of thin-calibre frankfurter-type sausages, *Fleischwirtschaft*, 65, 895, 1985.
24. **Daun, H.,** Interaction of wood smoke components and foods, *Food Technol.*, 33(5), 66, 1979.

Chapter 7

THE ANTIOXIDATIVE PROPERTIES OF WOOD SMOKE

I. INTRODUCTION

It has been known and proven for quite some time that certain levels and types of specific wood-smoke fractions have general antioxidative properties[1-11] and thus, from a practical standpoint, smoke can be used to retard lipid-associated rancidity in a wide variety of smoked foods.[12,13]

Because of the above facts, there has been interest in attempting to fractionate wood smoke in an effort to find the most efficient fraction relative to antioxidant properties. From a food additive standpoint, this could be of major significance since an antioxidant derived from the smoking of wood would be classified as a natural product. However, from a practical standpoint, these effective fractions also usually possess characteristic wood-smoke odors and thus, their use in certain bland foods would be somewhat limited.

For example, it has been reported[24] that if wood smoke is fractionated into acid, basic, and neutral portions, the neutral fraction which contains most of the phenols showed the best antioxidative properties. In contrast, the acidic portion had minor antioxidative properties while the basic fraction actually promoted lipid oxidation.

II. REACTIVE COMPOUNDS

It has been clearly demonstrated that phenols are the primary antioxidant-related compounds associated with wood smoke.[1-4] This is quite interesting in light of the fact that the most commonly used synthetic antioxidants used in the U.S. are also phenolic in nature, namely butylated hydroxyanisole (BHA) and butylated hydroxytoluene (BHT). In addition, it appears that high-boiling phenols such as 2,6-dimethoxyphenol, 2,6-dimethoxy-4-methylphenol, and 2,6-dimethoxy-4-ethylphenol are the key compounds responsible for antioxidative properties whereas low-boiling phenols only demonstrate weak antioxidant properties.[2,4]

The antioxidative properties of the phenolic derivatives associated with smoke as summarized in Table 1 have been demonstrated, but in all probability any phenolic that can produce a quinid structure will demonstrate some degree of functionality.

III. SMOKE SOURCE

Most of the studies performed in this area utilized liquid smoke, and for products of this type, levels of from 0.2 to 2% were found to be effective[8] and it then becomes apparent that smoke production, and thus, resulting composition, must be of importance. This represents another area of research that deserves further attention. However, the utilization of high levels of liquid smoke strictly for its antioxidative property can cause problems since it has been shown that at 10% addition, liquid smoke was actually prooxidative.[10] The reason for this observation was not addressed.

Research already conducted has found that smoldering smoke generation results in strong antioxidative properties.[5] Also, it has been demonstrated that the particle phase of smoke is more effective than the vapor phase.[2,7] Thus, with factors such as these in mind, it should be possible to generate smoke that has maximum antioxidant properties.

Table 1
SMOKE-RELATED PHENOLIC COMPOUNDS WITH PROVEN ANTIOXIDATIVE PROPERTIES

Pyrocatechol
Hydroquinone
Guaiacol
Eugenol
Isoeugenol
Vanillin
Salicylaldehyde
2-Hydroxybenzoic acid
4-Hydroxybenzoic acid

Adapted from Toth, L. and Potthast, K., *Adv. Food Res.*, 29, 87, 1984.

REFERENCES

1. **Kurko, B. I.,** Antioxidative properties of curing smoke components, *Myasn. Ind.*, 30(3), 19, 1959.
2. **Tilgner, D. J., Sikorski, Z. E., Daun, H. and Rudnikl, A.,** Antioxidant activity of the dispersed and dispersing phases of curing smoke, *XX-Lecie Politech. Gdansk. Sesja Nauk.*, 45, 344, 1965.
3. **Tilgner, D. J., Daun, H., and Rudnicki, A.,** Antioxidant activity of phenolic compounds present in curing smoke, *Rocz. Inst. Przem. Mieschego, Tluszczowego,* 4, 15, 1967.
4. **Daun, H.,** Antioxidative properties of selected smoked meat products, *15th Eur. Meet. Meat Res. Workers,* Helsinki, Finland, 1969.
5. **Radecki, A., Bednarek, R., Boltrukiewicz, J., Cocha, R., Rapicki, D., Grzybowski, J., Halkiewicz, J., and Lumparczyk, H.,** Analysis of industrial preparation of curing smoke. I. Separation and identification of higher phenol fractions, *Bromatol. Chem. Tokyskol.*, 8, 179, 1975.
6. **Tilgner, D. J. and Daun, H.,** Antioxidative and sensory properties of curing smokes obtained by three basic smoke generation methods, *Lebensm. Wiss. Technol.*, 3, 77, 1970.
7. **Daun, H. and Tilgner, D. J.,** The influence of generation parameters upon antioxidative properties of wood smoke, *Acta Aliment. Pol.*, 3, 231, 1977.
8. **Watts, B. M. and Faulkner, M.,** Antioxidative effects of liquid smokes, *Food Technol.*, 8, 158, 1954.
9. **Grabowska, J.,** Effect of polyhydroxyphenols on antioxidative activity of wood smoke, *Rocz. Panstw. Zakl. Hig.*, 22, 345, 1971.
10. **Chomiak, D. and Goryn, A.,** Course of lard antioxidation in the presence of liquid smoke, *Rocz. Inst. Przem. Mieschego Tluszczowego,* 13, 93, 1976.
11. **Chomiak, D. and Goryn, A.,** The inhibiting effect of smoke flavoring upon the autoxidation of lard, *Acta Aliment. Pol.*, 3, 223, 1977.
12. **Lea, C. H.,** The effect of smoking and influence of atmospheric humidity on keeping properties of bacon, *J. Soc. Chem. Ind.*, 52, 57T, 1933.
13. **White, W. H.,** Rancidity in pork fat after frozen storage and conversion to bacon, *Can. J. Res.*, 19D, 96, 1941.
14. **White, W. H.,** Smoked meats. II. Development of rancidity in smoked and unsmoked Wittshire bacon during storage, *Can. J. Res.*, 22F, 97, 1944.
15. **Smith, F. H., Brady, D. E., and Comstock, R. F.,** Rancidity of bacon, effect of antioxidant, *Ind. Eng. Chem.*, 37, 1208, 1945.
16. **Jensen, L. B.,** *Microbiology of Meats,* The Garrard Publishing, Champaign, Ill., 1945.
17. **Grant, G. A. and White, W. H.,** Keeping quality enhanced by curing and smoking poultry, *Food Can.*, 9(5), 14, 1949.

18. **Banks, A.,** Factors affecting the oxidation of the fat of dehydrated herrings, *J. Sci. Food Agric.*, 1, 28, 1950.
19. **Gaddis, A. M.,** Effect of pure salt on the oxidation of bacon in freezer storage, *Food Technol.*, 6, 294, 1952.
20. **Shevchenko, M. G. and Lapshin, I. I.,** The effect of smoking dried fish products on the stability of lipids to oxidation, *Rybn . Khoz.*, 11, 74, 1975.
21. **Woolfe, M. L.,** The effect of smoking and drying on the lipids of West African herring, *J. Food Technol.*, 10, 515, 1975.
22. **Goncharov, A. M. and Khamadi, L. A.,** Changes in fatty acid composition of lipids of catfish during hot smoking, *Rybn. Khoz.*, 3, 54, 1979.
23. **Kosev, D., Vasilev, K., and Kostov, L.,** Variation in the contents of volatile fatty acids induced by technological processes during the manufacture of smoked ocean mackeral, *Nauchni Tr. Vissh Inst. Khranit. Vkusova Prom.*, 26, 25, 1979.
24. **Toth, L. and Potthast, K.,** Chemical aspects of the smoking of meat and meat products, *Adv. Food Res.*, 29, 87, 1984.

Chapter 8

SMOKE AND FOOD COLOR

I. INTRODUCTION

One of the other functional properties of wood smoke is to impart color to foods. Color, in turn, is important since with most foods it is the first sensory factor, followed by aroma, flavor, and texture, that governs food acceptability. Therefore, it is quite important to understand the factors that govern color formation in smoked foods; however, unfortunately not all of the details are known.

Tyszkiewicz and Panasik[18] recently objectively measured the surface color and calculated the dominant wavelength, brightness, and color saturation of smoked sausage and reported that these values correlate closely with subjective panel evaluation. Thus, it would appear that people can have strong preferences for smoked food color.

From a simplistic standpoint, one could assume that smoked-food color results from the deposit of particles in the disperse phase of smoke directly on a food surface. Except in rather primitive smoking operations, this is usually not the case. However, research would indicate that there is more involved than just physical adsorption of particulate matter,[1-5] since deposition of smoke particles on an organic surface results in a darker color than the same particles deposited on an inert surface. This would indicate that a definite chemical reaction has occurred.

II. COMPOUNDS INVOLVED

A. Carbonyls

Research would indicate that the chemical reaction mentioned above involves the interaction of carbonyls in the smoke vapor phase with amino groups, derived from protein, on food surfaces.[6-12] In effect, this represents a series of nonenzymatic reactions similar to the Maillard reaction. In general, because of the amounts present, the carbonyls shown in Table 1 are thought to be major contributors to color formation in smoked meats.

However, based on smoking conditions, some may be more important than others. Of the compounds listed in Table 1, the most reactive carbonyls are formaldehyde, acetol, glycolic aldehyde, and methyglyoxal, with the latter two producing the most color.[1] The overall role of formaldehyde relative to degree of color formed is not clearly understood, since although the compound is very reactive with amino groups, little color seems to form.[5]

The fact that protein amino groups are involved has been clearly demonstrated by Ruiter[1] who used chemically modified collagen, whereby its amino groups were converted to hydroxy groups. No color developed during the smoking of hydroxy-containing collagen, whereas with normal collagen, extensive color formation resulted.

B. Phenols

Certain phenols associated with the vapor phase of smoke are also thought to contribute to color formation in smoked foods. However, they have to have a relatively high molecular weight[13] so as to have a sufficient number of hydroxyl groups to cross-link protein at numerous sites through hydrogen bonding. Two phenols that are thought to react in this manner are coniferylaldehyde and sinapaldehyde.[10] However, in most situations the amounts of these two polyphenols is rather low, especially compared to the amounts of carbonyls present.

Simple phenols are not thought to react to form color compounds but instead are absorbed by fat. In addition, it has been shown that simple phenols do not react with amino groups.[12]

Table 1
CARBONYLS ASSOCIATED WITH THE FORMATION OF COLOR IN SMOKED FOODS

Formaldehyde
Glycolic aldehyde
Glyoxal
Acetone
Hydroxyacetone
Methylglyoxal
Diacetyl
Furfural

Adapted from Ruiter, A., *Food Technol.*, 33(5), 54, 1979.

Table 2
MODEL SYSTEM REACTANTS THAT CONTRIBUTE TO COLOR

	Ref.
Glycolic aldehyde + aminoethanol → 1-hydroxyethyl-3-hydroxymethyl-2-pyrrolaldehyde	2
Glycolic aldehyde + aminoethanol + formaldehyde → 3-unsubstituted-1-hydroxymethyl-2-pyrrolaldehyde	3
Xylose + aminoethanol + formaldehyde → 3-unsubstituted-1-hydroxymethyl-2-pyrrolaldehyde	3
Methylglyoxal + methylamine → 1,5-dimethyl-4-hydroxy-2-pyrrolealdehyde	14
Dihydroxyacetone + methylamine → 1,5-dimethyl-4-hydroxy-2-pyrrolealdehyde	15

III. REACTION MECHANISMS

Since carbonyl-amine reactions can result in the formation of numerous compounds, some researchers have attempted to identify pigmented compounds from relatively simple model systems. Some of the model systems and resulting compounds that contribute to color in smoked foods are summarized in Table 2. As can be seen, several of the identified end products can be formed from different starting materials, thus increasing the amount of end product.

In comparing these end products with those obtained between aldose-amine reactions, as typified by the Maillard reaction, Ruiter[5] concluded that the basic mechanism is the same for the formation of color compounds in meat whereby there is a formation of conjugated unsaturated compounds which attach to hydroxy groups, with the resulting products condensing into brown-color compounds.

The only difference between the Maillard reaction and the smoke reaction is that in the formation of smoke color, the initial carbohydrate degradation occurs in the smoke generator and the resulting reactive compounds are brought into direct contact with reactive amino groups at the food surface without further rearrangement, whereas in the Maillard reaction a significant amount of rearrangement occurs via dealdolization and aldolization of fragments before the final reaction takes place. Undoubtedly, numerous compounds are formed during the formation of smoke color, but as with investigating the Maillard reaction, numerous intermediates, because of their complexity or lack of stability, await identification. To date, only the general scheme has been identified with the actual identification of all compounds involved being a long way off.

IV. FACTORS INFLUENCING RATE OF COLOR FORMATION

A. Concentration of Reactants

Obviously, a certain level of reactants is necessary before color will develop. In a model system situation, Ruiter[2] reported that maximum color formation required 2 to 3 mol of carbonyl per mole of amino compound. Thus, to maximize color formation, one should maximize carbonyl formation during wood degradation. Obviously, the wood composition and its moisture content will influence the level of carbonyls generated. Other factors to consider would be temperature of combustion and the presence of oxygen. A fairly detailed study considering all of these factors would be required to completely predict conditions for maximum carbonyl generation.

B. Temperature

As with the Maillard reaction, temperature is another factor to consider, with color developing faster as the product surface temperature increases. Thus, one would obviously produce more surface color formation in hot smoking vs. cold smoking, for example.

C. Moisture Level

Again, as with the Maillard reaction, product moisture content will influence the degree of color formation. If the reactant surface is too moist, the reaction will proceed slowly, while if the surface is too dry, the carbonyls from the smoke will not be absorbed. It appears that the maximum color formation on a smoked food surface occurs when the surface moisture content is between 6 and 10%.[16]

D. Method of Smoke Application

As one would expect, the way, and thus probably the amount, in which a food product is exposed to smoke can influence the resulting degree of color. This fact can perhaps best be appreciated by reviewing data generated by Sink and Hsu.[17] They evaluated various color properties of frankfurters that had been treated with four different types of smoke sources, plus a no-smoke control. The smokes included a conventional solid smoke aerosol, liquid smoke aerosol, liquid smoke external dip, and liquid smoke internal mix. The specific color properties measured included the pigment ratio (cured/total), reflectance and extinction ratios, percent pigment oxidation, and percent pigment ring (porphyrin) destruction. From their data, they concluded that liquid smoke dip resulted in a lighter color than the other methods, while the most stable color was observed for liquid smoke mix, with the least stable color attributed to the liquid smoke applied by aerosol.

Small changes in color were noted over a 3-week storage period for all variables. This color change did not result in changes in the values obtained for reflection and extinction ratios or percent pigment oxidation. However, the percent porphyrin ring of myoglobin destruction did continue to increase with storage for all treatments.

V. MISCELLANEOUS METHODS OF SMOKE COLORING

A. Steam-Generated Smoke

Several alternate procedures, some of them which have only been evaluated briefly on an experimental basis, are available to give certain foods a smoke-processed color. One such method involves the passing of superheated steam through wood shavings, thus resulting in a dense smoke which permits the adsorption of a highly concentrated liquid onto the food surface.[5] The process is fairly rapid and thus, a strong smoky flavor does not develop. In addition, there does not appear to be appreciable amounts of PAH associated with the process.

B. Electrostatic Deposition

Another approach involves the electrostatic deposition of wood smoke particles directly onto food surfaces. However, this procedure is not strongly encouraged for several reasons, including that the equipment is expensive and can pose a danger to food processors; strong off-flavors can be introduced; and PAH levels can be quite high.

C. Dyes

Commercial use of certain artificial azo dyes is another possibility. In the case of kippered herring marketed in the United Kingdom, the dye, Brown FK, has been used. The designation FK stands "for kippers". Other synthetic dye blends have been evaluated, but do not seem to be as effective at coloring as is Brown FK.

The natural dye annatto has also been evaluated, but it preferentially absorbs into fatty tissue at the expense of not coloring protein portions, thus resulting in uneven coloring.

D. Infared Radiation

An approach proposed by Ruiter[5] involves spraying a concentrated solution of the most active carbonyl compounds onto the surface of the food in question and then exposing the product to infrared radiation. This would certainly be an interesting approach, but many of the details remain to be investigated.

REFERENCES

1. **Ruiter, A.,** The coloring of protein surfaces by the action of smoke, *Lebensm. Wiss. Technol.*, 3, 98, 1970.
2. **Ruiter, A.,** The browning reaction of glycolic aldehyde with aminoethanol, *Lebensm. Wiss. Technol.*, 4, 38, 1971.
3. **Ruiter, A.,** The browning reaction of glycolic aldehyde with aminoethanol. II. Effect of formaldehyde, *Lebensm. Wiss. Technol.*, 5, 137, 1972.
4. **Ruiter, A.,** The browning reaction of glycolic aldehyde with aminoethanol. III. Renewed investigations on the structure of a reaction product and comparison with aldose-aminoethanol browning, *Lebensm. Wiss. Technol.*, 6, 142, 1973.
5. **Ruiter, A.,** Color of smoked foods, *Food Technol.*, 33(5), 54, 1979.
6. **Ziemba, L.,** Noneyzymatic browning reactions in smoked foods, *J. Nutr. Diet.*, 4, 122, 1967.
7. **Kersken, H.,** Hot smoking, *Fleischwirtschaft*, 54, 1721, 1974.
8. **Coretti, L.,** The color of smoked meat, *Fleischwirtschaft*, 55, 1504, 1975.
9. **Gilbert, J. and Knowles, M. E.,** The chemistry of smoked foods: a review, *J. Food Technol.*, 10, 245, 1975.
10. **Chen, L. B. and Issenberg, P.,** Interaction of wood smoke components with ε-amino groups in proteins, *J. Agric. Food Chem.*, 20, 1113, 1972.
11. **Ziemba, L.,** Coloration of the surface of foodstuffs in the process of smoking, *Przem. Spozyw.*, 20(11), 34, 1966.
12. **Ziemba, L.,** The filter paper disc test for evaluating the smoke color development and stability, *Przem. Spozyw.*, 21(7), 25, 1967.
13. **Caurie, M., Lee, T. C., Salomon, M., and Chichester, C. O.,** Hot smoke fish curing, *J. Natl. Sci. Counc. Sri Lanka*, 2, 77, 1974.
14. **Wieland, T. and Severin, T.,** Studies on the Maillard Reaction, *Z. Lebensm. Unters. Forsch.*, 153, 202, 1973.
15. **Manniger, G. and Severin, T.,** Studies on the Maillard Reaction, *Z. Lebensm. Unters. Forsch.*, 161, 45, 1976.
16. **Daun, H.,** Interaction of wood smoke components and foods. *Food Technol.*, 33(5), 66, 1979.
17. **Sink, J. D. and Hsu, L. A.,** Chemical effects of smoke processing on frankfurter quality and palatability characteristics, *Meat Sci.*, 3, 247, 1979.
18. **Tyskiewicz, S. and Panasik, M.,** Formation of surface color of sausages in the process of high-temperature smoking, *Proc. Eur. Meet. Meat Res. Workers*, 31, 533, 1985.

Chapter 9

ANTIMICROBIAL PROPERTIES OF SMOKE

I. INTRODUCTION

Another function of wood smoke, or at least certain fractions of it, is its potential antimicrobial properties. The word potential is used since there appears to still be questions in the literature as to how significant this property may be. This primarily arises from the fact that most in vivo studies attempting to evaluate smoke component effectiveness in retarding or preventing microbial growth that utilize traditional smoking procedures have compounding variables such as temperature and water activity that are inherent in their experimental design. Thus, one could argue that the true effect of smoke as an antimicrobial agent may be better evaluated by utilizing liquid smoke preparations.

Another area of confusion in the literature relative to the potential antimicrobial role of smoke is the fact that numerous strains of bacteria, mold, and yeasts can be naturally present and thus, one has to look at both total microbial counts as well as be aware that certain strains within each species may be more or less sensitive or resistant to smoke or some of its components.

In addition, the question as to which chemical fraction within wood smoke, or even to be more specific, which compound is primarily responsible for any observed antimicrobial properties. Certainly, individual smoke compounds can be evaluated in model systems, but it becomes difficult to effectively evaluate any synergistic effects of smoke components, especially in light of the number of compounds identified in smoke.

In the future, the concept that certain smoke components possess antimicrobial properties will probably be reevaluated since compounds of this type can be considered to be derived from natural sources and if certain individual compounds or combination of these compounds were found to be effective at low concentrations so as to not possess a characteristic smoky flavor, they could be used as substitutes for a wide range of currently used synthetic compounds.

The interaction of smoke components with microbes has seen two main areas of investigation, one deals with the reduction of microbial counts during food smoking and the other has primarily been concerned with extending the shelf life of smoked foods. Most of the research has been associated with bacterial growth, with much less work focusing on molds, and even less with yeasts. At this point, the discussion will be expanded on the three primary forms of microbes investigated.

II. ANTIBACTERIAL ASPECTS

As mentioned above, some researchers have concluded that wood smoke is not an effective bacteriostatic agent.[1-3] For example, Houben[1] evaluated the addition of two commercial liquid smoke preparations at levels of 0, 0.25, 0.5, 1.0, and 2% in a pork suspension at pH 6.3 that had been inoculated with individual strains of bacteria including *Bacillus cereus, B. subtilis, Escherichia coli, Lactobacillus* spp. 319, *Microbacterium* ssp C_2, *Micrococcus* spp. 1C, Staphyloccus aureus, and *Streptococcus faecium* E_{15} and E_{20}. Incubation was for 14 days at 18°C. Based on growth results, the authors concluded that the smokes had no significant effect on the bacteriological stability of pork.

Stankiewicz-Berger and Kitzman[3] also evaluated the bacteriostatic activity of a commercial liquid smoke preparation that was composed of a phenolic fraction containing 24.3% phenol and homologues, 29.5% guaiacol and 37% syringol; an acidic fraction having 30% acetic acid and homologues, 25% cyclotene and homologues and 40% phenols; and a neutral

Table 1
INFLUENCE OF LIQUID SMOKE CONCENTRATE ADDITIONS ON SAUSAGE QUALITY

Concentrate (g/kg)	Keeping time[a]	Bacterial reduction[b]
0	1.00	1.00
10	1.08	3.18
20	1.18	3.89
30	1.30	4.44
40	1.45	4.92
50	1.63	5.35
80	2.64	6.47
100	4.46	7.14
120	14.42	7.77

[a] Ratio of keeping time for samples with added concentrate to the keeping time for control samples not containing concentrate.
[b] Ratio of bacterial counts in control samples to the bacterial counts in samples with added concentrate.

Adapted from Olsen, C. Z., *Proc. Eur. Meet. Meat Res. Workers*, No. 22, F7:1, 1976.

fraction containing 30% ketones and other oxy-compounds and 30% hydrocarbons. Bacterial strains evaluated included *E. coli*, *Clostridium sporogenes*, and *S. faecalis*. They found that none of the strains were inhibited by smoke concentrations of less than 200 ppm. Since the smoke preparation had a smoke threshold of approximately 20 ppm, they concluded that this liquid smoke preparation would have limited use in foods not associated with a typical smoked flavor.

In a short-time incubation study for 24 hr at 37°C, Oluski and Zivanovic[2] concluded that a liquid smoke preparation, when used at levels up to 2%, did not have bacteriocidal or bacteriostatic properties against *B. subtillis*, *B. cereus*, *Sarcina subflava*, *S. lutea*, *S. aureus*, *S. epidermidis*, or *M. flavus*.

In contrast to the above studies, numerous researchers have reported both specific and generalized antimicrobial properties associated with smoke.[4-23] For example, Olsen[4] demonstrated that a liquid smoke preparation used at a concentration of 6.5 g/kg prolonged the lag time of *S. aureus*, which was at a concentration of 10^5/mℓ, by 4 days at 30°C. When 9.8 g of the smoke preparation was used, lag time was prolonged to 14 days. He also evaluated its effectiveness in ground pork loin and found that a concentration as low as 3.5 g/kg of meat significantly retarded bacterial lag time. In the manufacture of fermented sausage using a lactic acid starter, Donnelly et al.[8] also demonstrated that the initial lag phase was delayed by the addition of liquid smoke. Kozlowski et al.[5] also reported that a liquid smoke extract was effective in both prolonging the shelf life and reducing bacterial counts in a sausage product. As seen in Table 1, the addition of only 0.01 g/kg of extract reduced counts by a factor of three, while higher levels were even more efficient.

Liquid smoke in combination with sodium chloride was found to be effective in preventing growth and toxin production associated with *Clostridium botulinim* types A and E spores in several fish species that were stored at 25°C for 7 and 14 days.[14] When liquid smoke was used, salt level could effectively be reduced from 4.6 to 2.8% and still provide protection for 7 days. This points out an interesting concept that has received little attention, namely synergistic interaction among preservatives. It would appear that smoke fractions could be

Table 2
INFLUENCE OF SMOKE GENERATION TEMPERATURE ON BACTERIAL DELAY OF GROWTH (DAYS)

Organism	Smoke generation temperature (°C)	Smoke concentration (ppm) vs. days of delayed growth
S. aureus	350	500/2.5 days
	400	500/1 day
	450	500/1 day
	500	500/1 day
E. coli	350	1250/1 day
	400	1250/1 day
	450	1250/1 day
	500	1250/1 day

Adapted from Fretheim, K., Granum, P. E., and Vold, E., *J. Food Sci.*, 45, 999, 1980.

effectively used with other traditional chemical preservatives, including salt, and lower the total amount of preservative used. This could be especially beneficial in preserved products normally containing significant amounts of salt, especially in light of recommendations to reduce our dietary salt intake.

Houben[15] conducted an interesting study whereby he evaluated the effectiveness of five commercial smoke preparations against bacteria that are of significance in meat products, namely *B. cereus, B. subtillis, E. coli, Micrococcus* sp., *S. aureus, S. faecalis*, and *Lactobacillus* sp. Two of the bacteria showed very little bacteriostatic activity, two were moderately active, and one was strongly bacteriostatic in the presence of smoke. This would perhaps indicate that possibly wood source or method of smoke generation can have a sigificant influence on effectiveness and thus, perhaps effectiveness could be substantially increased by modifying the smoking process.

For example, Fretheim et al.[17] generated smoke at temperatures ranging from 350 to 500°C and evaluated its effectiveness against certain bacteria. As seen in Table 2, growth at 30°C was delayed by the incorporation of smoke and there were differences in sensitivity between the two bacterial species evaluated with *E. coli* being more resistant than *S. aureus*. They also pointed out that, at least with *S. aureus*, not only was lag phase influenced but its overall growth rate was lower in the presence of smoke. This study was only performed in a model system, but if the same trends hold true for certain food systems, smoke could be used to not only retard certain bacterial growths, but to also lower total counts with time.

In contrast, Shigehisa et al.[29] found that when boneless pork loins were intentionally inoculated with *E. coli, Salmonella typhimurium*, or a mixed strain of *S. aureus* forming enterotoxins A, B, and C and then smoked, *E. coli* and *S. typhimurium* counts were greatly reduced but *S. aureus* was much more resistant.

Likewise, Houben[15] found that bacterial strains differed significantly in their sensitivity to smoke. *Bacillus* spp. and *Microbacterium* sp. were the most sensitive while *E. coli* and Lactobacillaceae were most resistant. This could account for some of the apparent discrepancies that appear in the literature relative to the effectiveness of smoke as a bacterial inhibitor.

Olsen[16] has evaluated the effectiveness of various smoke fractions against *S. aureus*. The total smoke condensate as well as the terpene, phenol, and three subphenol fractions were obtained and evaluated. One of the phenol subfractions and the terpene fractions were found to be ineffective while another of the phenol subfractions prolonged the lag phase. The total

Table 3
CONCENTRATIONS (ppm) OF PHENOLS CAUSING A 50% GROWTH REDUCTION OF DIFFERENT YEAST STRAINS

Yeast/type	Phenol	*p*-Cresol
Pichia (fermentative)	900	400
Saccharomyces sp. (fermentative)	900	ND
Rhodotorula sp. (oxidative)	500	ND
Rh. rubra (oxidative)	1000	200
Torulopsis sp. (oxidative)	460	ND

Note: ND = not determined.

Adapted from Kwasniewska, K. and Kaiser, K. L. E., *Bull. Environ. Contam. Toxicol.*, 31, 188, 1983.

phenol fraction and the third phenol subfraction did not influence lag phase but did reduce total bacteria concentration. The total smoke condensate at 147 and 222 ppm prolonged the lag phase 4 and 14 days, respectively. This evidence would indicate that certain phenols may be primarily responsible, however no extensive study investigating the role of other smoke fractions has been reported. In contrast, Nielsen and Pendersen[22] concluded that in smoked salmon, *C. botulinum* types B and E were inhibited not by phenols but by formaldehyde. Likewise Zaitsev et al.[23] implicated formaldehyde along with phenols and organic acids as being responsible for bacterial inhibition. Thus, it would appear that the compounds responsible for bacterial inhibition have not been totally identified and species variation is apparent.

III. ANTIFUNGAL AND YEAST ASPECTS

In general, it would appear that molds and yeasts are less sensitive to smoke than bacteria.[2,10,24] Also, as is the case with bacteria, species differences can be significant. For example, Kersken[25] evaluated that fungistatic effect of smoke on six species of mold common to cured meat products and reported that *Cladosporium herbarium* was quite sensitive to smoke whereas the three aspergillus species were resistant, and perhaps smoke even acted as a stimulant for the growth of these latter molds. He also evaluated the influence of smoke generation temperature (300, 350, or 400°C) and found that fungistatic effect increased with temperature of smoke production but no attempts were made to identify the specific compounds responsible.

Arseculeratine et al.[26] have reported that the smoke treatment of copra kernels inhibited aflatoxin production by *Aspergillus flavus* but did not significantly inhibit mycelial growth. The same observations were made by Alverez-Barrea et al.[27] in the case of fermented sausages.

Recently, Kwasniewska and Kaiser[28] evaluated the effectiveness of various phenols, including the compounds phenol and *p*-cresol which can be major components of wood smoke, on a series of fermentative and oxidative yeasts. As seen in Table 3, *p*-cresol was more inhibitory than phenol for all yeast strains evaluated. Because of the generally higher levels required for effectiveness against yeasts, relatively little has been done in this area.

REFERENCES

1. **Houben, J. H.**, The bacteriostatic properties of smoke preparations. II. Inhibitory activity in meat suspensions, *Voedingsmiddelen Technol.*, 8(15), 5, 1975.
2. **Oluski, A. and Zivanovic, R.**, Bacteriocidal activity of a Yugoslav smoke preparation, *Tehnol. Mesa*, 17(12), 374, 1976.
3. **Stankiewicz-Berger, H. and Kirman, P.**, Bacteriostatic activity of refined liquid smoke, *Acta Aliment. Pol.*, 5(4), 291, 1979.
4. **Olsen, C. Z.**, Chemical composition and application of smoke flavor, *Proc. Eur. Meet. Meat Res. Workers*, No. 22, F7:1, 1976.
5. **Kozlosdki, Z. P., Simon, J., and Wilimowski, M.**, Investigations on the chemical, toxicological, and biological properties of the Polish smoke extract, *Proc. Eur. Meet. Meat Res. Workers*, 15, 516, 1969.
6. **El-Ashwah, E. T., El-Magoli, S. B., and Ibrahim, M. M.**, Studies on curing, ageing and smoking of camel meat. II. Microbiological and sensory evaluation, *Proc. Eur. Meet. Meat Res. Workers*, No. 27, Vol. 1., C:9, 238, 1981.
7. **Smith, J. L., Palumbo, S. A., Kissinger, J. C., and Huhtanen, C. N.**, Survival of *Salmonella dublin* and *Salmonella typhimurium* in Lebanon bologna, *J. Milk Food Technol.*, 38, 150, 1975.
8. **Malle, P., Eb. P., and Tailliez, R.**, Study of the antiseptic action of smoke on *Escherichia coli*, *Ind. Aliment. Agric.*, 98(3), 175, 1981.
9. **Kossakowska, A., Wojton, B., and Widenska, T.**, Behavior of *Salmonella* organisms during sausage production, *Bull. Vet. Inst. Pulawy*, 18(3)/4), 57, 1974.
10. **Barylko-Pikielna, N.**, Contribution of smoke compounds to sensory, bacteriostatic and antioxidative effects in smoked foods, *Pure Appl. Chem.*, 49, 1667, 1977.
11. **Mourey, A. and Bracquart, P.**, Bacterial contamination of cured meat products, *Med. Nutr.*, 15, 431, 1979.
12. **Donnelly, L. S., Ziegler, G. R., and Action, J. C.**, Effect of liquid smoke on the growth of lactic acid starter cultures used to manufacture fermented sausage, *J. Food Sci.*, 47, 2074, 1982.
13. **Gibbons, N. E., Rose, D., and Hopkins, J. W.**, Bactericidal and drying effects of smoking on bacon, *Food Technol.*, 8, 155, 1954.
14. **Eklund, M. W., Pelroy, G. A., Paranjpye, R., Peterson, M. E., and Teeny, F. M.**, Inhibition of *Clostridium botulinum* types A and E toxin production by liquid smoke and NaCl in hot-process smoke-flavored fish, *J. Food Prot.*, 45, 935, 1982.
15. **Houben, J. H.**, Bacteriostatic properties of smoke preparations. I. Comparative studies on five preparations, *Voedingsmiddelen Technol.*, 7(41), 8, 1974.
16. **Olsen, C. Z.**, Smoke flavoring and its bacteriological and antioxidative effects, *Acta Aliment. Pol.*, 3(3), 313, 1977.
17. **Fretheim, K., Granum, P. E., and Vold, E.**, Influence of generation temperature on the chemical composition, antioxidative, and antimicrobial effects of wood smoke, *J. Food Sci.*, 45, 999, 1980.
18. **Jensen, L. B.**, Action of hardwood smoke on bacteria in cured meats, *Food Res.*, 8, 377, 1943.
19. **Jensen, L. B.**, *Microbiology of Meats*, 3rd ed., Garrard Publishing, Champaign, Ill., 1954, 199.
20. **Southcott, B. A. and Razzell, W. E.**, *Clostridium botulinum* control in cold-smoked salmon: a review, *J. Fish. Res. Board Can.*, 30, 631, 1973.
21. **Shewan, J. M.**, The biological stability of smoked and salted fish, *Chem. Ind.*, 501, 1949.
22. **Nielsen, S. F. and Pedersen, J.**, Studies on the occurrence and germination of *C. botulinum* on smoked salmon, in *Botulism*, Ingram, M. and Roberts, T. A. Eds., Chapman and Hall, London, 1967, 66.
23. **Zaitsev, V., Kizevetter, I., LaGunov, L., Makarova, T., Minder, L., and Podsevalov, V.**, *Fish Curing and Processing*, MIR, Moscow, 1969.
24. **Todorovic, M., Oluski, A., and Skrinijar, M.**, Fungistatic and fungicidal effects of potassium sorbate and smoke preparations on certain molds, *Tehnol. Mesa*, 20, 249, 1979.
25. **Kersken, H.**, Effects of smoke on molds, *Fleischwirtshaft*, 54, 1341, 1974.
26. **Arseculeratine, S. N., Samarajeewa, U., and Welianga, L. V.**, Inhibition of aflatoxin accumulation in smoked substrates, *J. Appl. Bacteriol.*, 41, 223, 1976.
27. **Alvarez-Barrea, V., Pearson, A. M., Price, J. F., Gray, J. I., and Aust, S. D.**, Some factors influencing aflatoxin production in fermented sausages, *J. Food Sci.*, 47, 1773, 1982.
28. **Kwasniewska, K. and Kaiser, K. L. E.**, Toxicities of selected phenols to fermentative and oxidative yeasts, *Bull. Environ. Contam. Toxicol.*, 31, 188, 1983.
29. **Shigehisa, T., Yamada, R., Taji, S., Shinagawa, K., and Sakaguchi, G.**, The bacteriological safety of unpasteurized cured meat products in response to the manufacturing processes, *Jpn. J. Vet. Sci.*, 47, 443, 1985.

Chapter 10

POTENTIAL HEALTH CONCERNS ASSOCIATED WITH SMOKE

I. INTRODUCTION

Up until now we have primarily talked about the positive aspects of wood smoke. However, in all fairness, we should also look at the reported negative aspects of wood smoke relative to human health. Since we are talking of humans, one must realize that moral and legal limitations make this an extremely difficult area to gather scientific evidence from well-designed studies. Therefore, mainly inferences have been made from animal studies and human death rate statistics. One can argue that humans have been consuming smoked foods for literally centuries and anyone would have an extremely difficult time in proving beyond a doubt that man has significantly threatened his life because of this practice.

However, based on the chemistry of certain compounds that can be formed in smoke, namely polycyclic aromatic hydrocarbons (PAH) and nitrosamines, a rational approach would be to minimize the ingestion of such compounds not especially by decreasing the amount of smoked foods consumed, but by reducing the levels of such compounds in our total commercial and home-processed food supply.

It should be noted that certain countries have established maximum limits for these compounds in commercial foods, but little if anything is done to educate the general public as to how these compounds can be reduced in home processed and prepared foods.

Human nature is unique in that a commercial food processing operation can quickly receive condemnation for marketing food that contains potentially harmful compounds, but no publicity or outrage is voiced to the individual or entire society who produces these same compounds in their homes. For example, PAH can be generated from various sources, and thus, can enter our food supply via several routes. PAH can be thermally produced by both high temperature pyrolysis and from the incomplete combustion of materials containing carbon and hydrogen. Various forms of air pollution can be a significant source of PAH. In addition, PAH are naturally associated with petroleum and coal tar products which can also be environmental pollutants, thus resulting in contaminated food supplies.

Health hazards associated with smoke, soot, and tar products were apparent long before the compounds responsible were known. For example, skin cancer was considered to be an occupational hazard of chimney sweeps centuries ago but the actual cause was not realized until the 1930s. Researchers discovered that PAH isolated from combustion residues were carcinogenic in animal tests.[40] The problem is compounded by the fact that apparently PAH is not acutely toxic and thus time is required for symptoms to develop.

Relative to the general thermal formation of PAH, it is generally thought that during the pyrolysis of organic matter the compound acetylene is a key intermediate, as outlined in Table 1. Since acetylene can be derived from many forms of organic starting materials, it is then not surprising that the occurrence of PAH is widespread. In fact, it has been shown, as seen in Table 2, that a relatively simple compound, such as glucose, can produce benzo[a]pyrene when pyrolyzed, and as structure complexity increases, so does the amount of PAH formed. Thus, in the heating or smoking of a food, which contains a vast number of potential reactants, PAH formation is a standard fact of life.

II. SMOKING METHODS VS. PAH CONCENTRATION

PAH is composed of multiple fuzed benzene rings and at least on a theoretical basis hundreds of such compounds can exist. To date, 40 to 60 such compounds have actually

Table 1
PYROSYNTHESIS OF BENZO[A]PYRENE

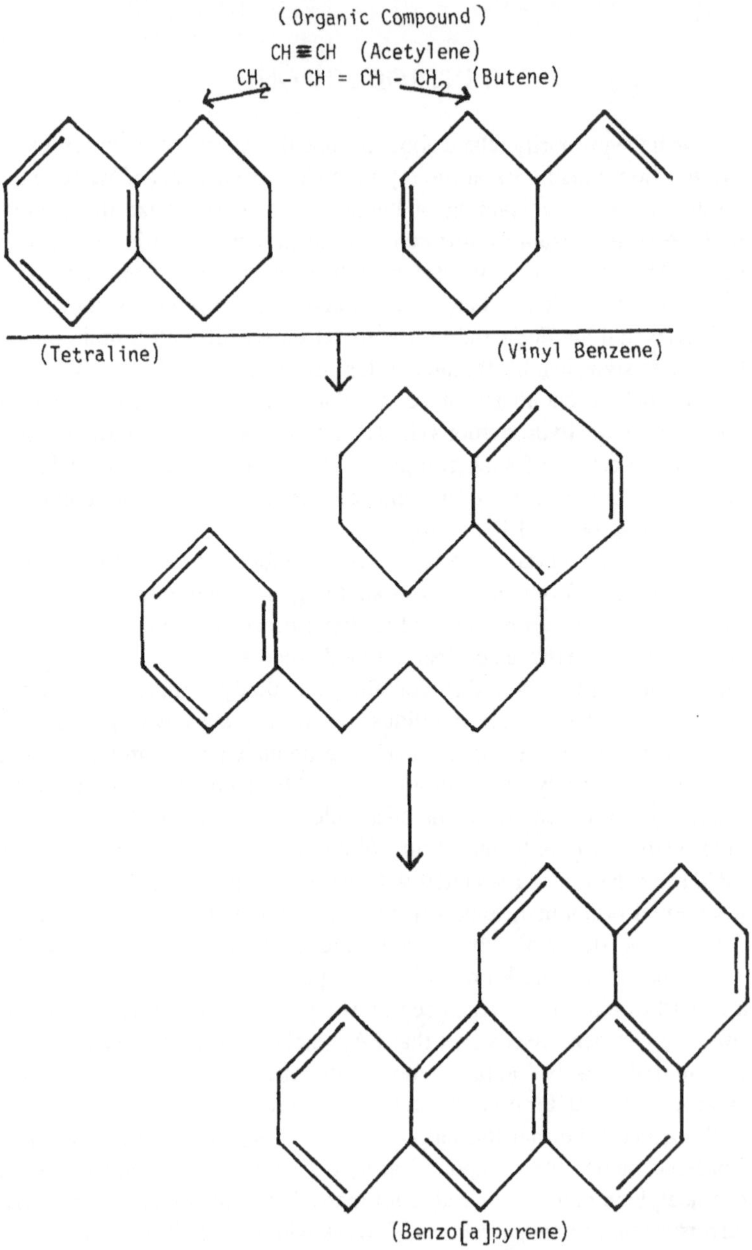

Adapted from Badger, G., Kimber, R., and Spotswood, T., *Nature*, 187, 663, 1960.

been identified although up to 100 have actually been detected. Some of the more common PAH are listed in Table 3.

Since it has been known for quite some time that some PAH act as potent chemical carcinogens (Table 4), their presence in food is of major concern. Smoked foods are a special area of concern since it has been well documented that smoke generation conditions can dramatically influence the level of resulting PAH in the smoke and consequently, in smoked foods.[1-21] At this point, certain of these studies will be discussed.

Table 2
BENZO[A]PYRENE GENERATION FROM THE PYROLYSIS (840°C) OF VARIOUS ORGANICS

Compound pyrolyzed	μg benzo[a]pyrene/kg pyrolyzate
Glucose	48
Stearic acid	1200
β-Sitosterol	3750

Adapted from Schmeltz, I. and Hoffman, D., *Carcinogenesis*, Vol. 1, Frendenttal, R. I. and Jones, P. W., Eds., Raven Press, New York, 1976.

Table 3
COMMON FOOD PAH

Indene	Fluoranthene
Methylindene	Benzo[a]anthracene
Ethylindene	Chrysene
Naphthalene	Benzo[b]fluoranthene
Methylnaphthalene	Benzo[k]fluoranthene
Dimethylnaphthalene	Benzo[j]fluoranthene
Ethylnaphthalene	1,2-Benzpyrene
Naphthylene	3,4-Benzyprene
Napthene	Perylene
Ethylnaphthalene	7,12-Dimethylbenz[a]anthracene
Methylnaphthylene	Methylchloroanthrene
Trimethylnaphthalene	Indenopyrene
Fluorene	Benzo[g,h,i]perylene
Methylfluorene	Anthranthrene
Phenanthrene	Dibenzo[a,c]anthracene
Anthracene	Dibenzo[a,j]anthracene
2-Methylanthracene	Dibenzo[a,h]anthracene
9-Methylanthracene	Coronene
Methylphenanthrene	3,4,9,10-Dibenzpyrene
Pyrene	Benzo[c]phenanthrene
Dimethylanthracene	Dibenzo[a,h]pyrene
Dimethylphenanthrene	Dibenzo[a,i]pyrene
1-Methylpyrene	Dibenzo[e,l]pyrene
11-H-benzo[a]fluorene	Cholanthrene

In the case of a frankfurter-type product, Simon et al.[1] demonstrated that the type of casing used during the smoking process could significantly influence benzo[a]pyrene concentrations in the resulting product. As seen in Table 5, sheep casing apparently permitted the passage of PAH into the product while the synthetic cellulose casing retarded PAH penetration. Also, as would be expected, increased smoking time increased product PAH concentrations. Apparently, the structure of the cellulose casing was such that the rather large molecular weight PAH could not effectively pass into the product. Thus, based on these data, the use of certain natural casings should be discouraged with smoked foods.

Rusz et al.[2] clearly demonstrated that electrostatic precipitators operating at a voltage of 10 kV and a temperature of 75 to 80°C were quite effective in removing PAH from smoke streams. Lenges[3] also promoted the use of electrostatic precipitators for PAH reduction.

Toth and Blaas[4] reported on numerous smoke-processing techniques that can significantly lower PAH levels. They noted a linear increase in benzo[a]pyrene concentration as smoke

Table 4
DEGREE OF CARCINOGENICITY AMONG CERTAIN PAH

Compound	Degree of carcinogenicity
7,12-Dimethylbenz[a]anthracene	+ + + +
3-Methylcholanthrene	+ + + +
Benzo[a]pyrene	+ + +
Dibenz[a,h]anthracene	+ + +
Dibenzo[a,i]pyrene	+ + +
Benzo[a]phenanthrene	+ + +
Benzo[y]anthracene	+
Indeno[1,2,3-c,d]pyrene	+
Benzo[b]fluoranthene	+
Pyrene	−
Benzo[a]fluorene	−

Note: + + + + extremely carcinogenic; + + + very carcinogenic; + + carcinogenic; + somewhat carcinogenic; − not carcinogenic.

Adapted from Toth, L. and Potthast, K., *Adv. Food Res.*, 29, 87, 1984.

Table 5
INFLUENCE OF CASING TYPE AND SMOKING TIME ON BENZO[A]PYRENE LEVELS (ppb) IN SMOKED FRANKFURTERS

	Smoking time (min)	
Type of casing	5	10
Sheep	4.0	13.0
Cellulose	1.6	1.6

Adapted from Simon, S., Rypinski, A. A., Tauber, F. W., Pencyla, R. M., and Westerberg, D. O., *J. Agric. Food Chem.*, 17, 1128, 1969.

generation temperature was increased from 400 to 1000°C. At the higher temperatures, primarily four- and five-membered PAH compounds were formed. Thus, it appears that smoke generation temperature is the key factor influencing PAH formation. They[4] demonstrated that when smoke generation temperature was 400°C, smoked meat had 0.4 ppb PAH, while if smoke generation temperature was reduced to 300°C, PAH levels in the meat fell to 0.09 ppb. Other significant factors resulting in reductions of smoke-generated PAH included cooling, which reduced PAH levels 25%, washing which resulted in a 20 to 30% reduction, increasing distance from the smoke generator — 40%, and filtering smoke for a 90% reduction. As would be expected, they[4] also reported that more PAH was deposited on food during hot smoking vs. cold smoking and a very high buildup of PAH during the dark smoking of food was observed due to soot buildup on the product surface. Raja et al.[18] and Potthast[19] have also clearly demonstrated that moistening the wood source during smoking effectively lowers smoke generation temperature, and thus, also PAH levels.

Table 6
INFLUENCE OF SMOKE GENERATION TEMPERATURE (°C) ON PAH FORMATION (μg/100 g OF WOOD)

Compound	Smoke generation temperature 400	600	800	1000
Benzo[a]pyrene	8	10	15	18
3,4-Benzofluoranthene	9	12	18	22
Pyrene	35	60	80	115
Fluoranthene	70	100	135	175
Total	122	182	248	330

Adapted from Lenges, J., *Rev. Ferment. Ind. Aliment.*, 27, 53, 1972.

A recent study has shown that at least in the case of benzo[a]pyrene, commercial smoking units represented by smoldering in open or closed systems, friction smoke, and steam smoke generation all produced frankfurters that had 0.1 to 0.7 ppb.[155]

Another study[156] has shown that the benzo[a]pyrene levels in European smoked meat products has changed very little over the 10-year period from 1975 to 1984 perhaps indicating that equipment design and operation is such that little more can be done to reduce or totally eliminate PAH from smoked meats. However, they did report that in the case of black-smoked products, approximately 40% of all samples exceeded the suggested 1 ppb benzo[a]pyrene level with one such product having 570 ppb present. Recent analysis of liquid smoke preparations have also found acceptable levels of benzo[a]pyrene present.[157]

The potential problem of only measuring benzo[a]pyrene content as influenced by smoke generation temperature as an indicator of total PAH is clearly shown in Table 6. It can be seen that several PAH compounds are present at significantly higher levels than benzo[a]pyrene, and at higher generation temperatures these compounds are actually formed at a faster rate than benzo[a]pyrene.

Similar results were obtained for alder chip smoke in that benzo[a]pyrene was found at a very low concentration (less than 1 ppb) whereas PAH derivatives of methylphenanthrenes were found at levels ranging from 3.1 to 43.1 ppm and isomeric forms of pentamethyl-phenanthrenes were present at levels ranging from 24.9 to 99.8 ppm.[154] Thus, if one applied the general rule that smoke or food containing less than 1 ppb benzo[a]pyrene is satisfactory, one is not absolutely sure as to its overall safety.

Toth and Blaas[5] also found that the total quantity of potentially harmful PAH in smoked meats was five to ten times the amount of benzo[a]pyrene detected. In light of this, the question arises if the German law permitting a maximum of 1 ppb benzo[a]pyrene in smoked foods is actually strict enough. Granted, the law is a good one, but perhaps it needs to be expanded to include other PAH.

With smoked poultry, Raja et al.[18] demonstrated that a restricted air supply and moistened wood shavings used as a smoke source produced a product lower in benzpyrene than smoking for the same time but with unlimited air and dry wood. Conditions such as these are usually not difficult to control in a commercial smoking operation, but one must wonder how many noncommercial smokers are operated with these parameters in mind. For example, Fritz and Soos[36] surveyed a wide range of commercially produced German smoked meats and found an average benzo[a]pyrene content of 0.43 μg/kg of product whereas artisanal smoked meats had an average of 0.76 μg/kg. In contrast, they found that Hungarian commercially smoked meats had an average of 0.60 μg/kg while Hungarian home smoked meats had 0.74 μg/kg.

They[36] also measured the benzo[a]pyrene level in smoked meats traditionally prepared by using soft wood as the smoke source by a Slovenian minority known to have a higher incidence of stomach cancer than the general population and these products were found to contain on the average 4.16 μg/kg.

As mentioned elsewhere, the filtration of smoke can significantly lower PAH concentration. For example, Steinig and Schreiber[11] evaluated the effectiveness of various types of metal (perforated sheets, steel wool, and expanded metal) and fiberglass filters. The fiberglass filters were found to be more effective, but more wood was needed to produce the equivalent amount of smoke as when no filter was used. Also, although more effective, the fiberglass filters could not be effectively cleaned to be reused whereas the metal filters could be reused upon cleaning.

Direct and indirect smoking conditions relative to resulting benzo[a]pyrene concentration in smoked fish was evaluated by Steinig.[9] It was found that the direct method gave levels greater than 1 ppb whereas indirect smoking produced products that had less than 1 ppb.

Toth and Blaas[5] examined benzo[a]pyrene levels in cold-, hot-, and black-smoked meats and found quite a range in levels with the black-smoked meats having the highest amounts. Traditional cold-smoked meats had an average of 0.6 ppb with a high of 1.6 ppb while hot-smoked products averaged 0.4 ppb with a high of 1.1 ppb. Since the cold-smoked meats were smoked for a longer period of time, they had higher levels than the hot-smoked products. In contrast, the black-smoked products ranged from 0.3 to 55 ppb.

Dikun et al.[20] demonstrated with fish products that using liquid smoke produced lower benzpyrene levels (1 to 1.2 μg/kg of fish) than by smoking using two commercial smoking units whose levels ranged from 1.6 to 39.0 μg/kg and averaged 9.7 μg/kg. Similar results with liquid smoke were reported by Ruchkovskii et al.[48] Thus, if product quality can be maintained, good quality liquid smoke can be used to minimize PAH.

Larsson[47] evaluted PAH types and amounts in Swedish commercially smoked fish as influenced by wood source. Either alder or spruce/juniper was used. Surprisingly, fish smoked over alder had a higher total PAH concentration (439 μg/kg of fish) than for spruce/juniper (271 μg/kg). However, there was a greater proportion of higher molecular weight PAH associated with the spruce/juniper wood source.

It is interesting to note that most commercial food smoking operations have routinely adopted many or all of the above steps to minimize PAH buildup in smoked foods, but the individual who is smoking foods for his own personal use usually smokes in a rather primitive manner, which in turn can lead to high PAH concentrations.

In light of the German maximum limit of 1 ppb 3,4-benzopyrene in smoked foods, Potthast[40] has summarized data obtained from several research groups on the influence of smoking method on resulting 3,4-benzopyrene content. A portion of his data is shown in Table 7, and as can be seen, overall, the method resulting in the lowest level is the wet smoke procedure. In contrast, either hot or cold dark smoking and the smoldering smoke method produced samples that had relatively high 3,4-benzopyrene amounts. Obviously, not all samples evaluated for each method had equal amounts of 3,4-benzopyrene as can be seen by the range of values found for each method, thus demonstrating that not only is the method of smoking important but the operator of the smoking unit also apparently plays an important role since these were commercial samples that were evaluated.

Potthast[7,15] has perhaps best summarized optimum smoke generation conditions required to minimize PAH. Since PAH generation is temperature dependent, he suggests generating smoke below 400°C, removing particulate matter through filtration or by means of a cold trap, wrapping products to be smoked in synthetic casings that are impermeable to PAH, using spatial separation of the product being smoked from the smoke source, and reducing the smoking time of products that have high surface-to-weight ratios.

Relative to filtration of smoke as a means of reducing PAH content, Potthast et al.[12] have

Table 7
INFLUENCE OF SMOKING METHOD ON RESULTING 3,4-BENZOPYRENE CONTENT IN COMMERCIAL SMOKED MEAT PRODUCTS

Smoking method	3,4-Benzopyrene (μg/kg)
Smoldering smoke	0—36.23
Wet smoke	0.09—0.77
Friction smoke	0.01—4.56
Hot smoke	
Lightly smoked	0—2.08
Dark smoked	0.22—36.23
Cold smoke	
Lightly smoked	0.10—3.16
Dark smoked	0.10—56.04

Adapted from Potthast, K., *Fleischwirtschaft,* 58, 38, 1978.

shown that the technique can reduce PAH levels up to 97% while cooling generated smoke can result in up to a 35% reduction. Also, the washing of smoke after generation can reduce PAH by 35%. Fedonin et al.[192] have also demonstrated the effectiveness of filtering smoke.

Another potential method for reducing the PAH level in smoked foods in through the use of smoke-flavored seasonings.[34] These seasonings were found to contain up to 20 μg/kg of benzopyrene, but when used at a maximum dose of 5 g of seasoning per kilogram of meat product, the resulting benzopyrene content was below the German 1 ppb standard.

Also to be added to the list is the use of hardwoods instead of softwoods for smoke generation, and in the case of smoking fish, removal of the skin since fish smoked without skin contained an average of 1 to 39 μ/kg of benzopyrene whereas the same fish smoked with the skin on contained 8.55 μg/kg.[55] Obviously, rapid hot smoking and black smoking result in high PAH levels[4] as does smoke generation with an open flame.

As discussed elsewhere, food lipid content is another factor that can contribute to PAH formation during smoking, and thus where possible, the lipid content of the food being smoked should be kept to a minimum. This appears to be especially true for smoked fish.

III. PAH IN WOOD SMOKE

Since PAH can be associated with wood smoke, numerous researchers have attempted to separate, identify, and quantitate such compounds. For example, Obiedzinski and Borys[23] detected 27 PAH while Potthast[17] identified 25 PAH compounds in smoke condensates. The major PAH in smoke appears to be phenanthrene[17,24] and closely related compounds. The same observation was noted earlier by Rhee and Bratzler,[22] and as can be seen in Table 8, whole smoke had approximately twice the PAH as the vapor phase thus again reinforcing the concept that the level of smoke particulate matter permitted to reach a food surface should be minimized.

The other major research area has been to evaluate PAH composition of liquid smoke since, in theory, techniques described in the previous section are available to reduce PAH levels. For example, as seen in Table 9, a tar recovered from a liquid smoke preparation

Table 8
PAH COMPOSITION IN WHOLE AND VAPOR PHASE WOOD SMOKE

Compound	Whole smoke	Vapor phase
Phenanthrene	51.5[a]	28.4
1,2-Benzanthracene	7.0	4.3
Fluoranthene	5.7	4.2
Pyrene	5.5	4.1
Anthracene	3.8	1.9
Chrysene	2.6	0.3
3,4-Benzopyrene	1.2	0.4
1,2-Benzopyrene	0.9	Trace
Total	78.2	43.6

[a] μg/4.5 kg of hard maple sawdust smoldered at 750 to 800°C.

Adapted from Rhee, K. S. and Bratzler, L. J., *J. Food Sci.*, 33, 626, 1968.

Table 9
PAH COMPOSITION OF A TAR RECOVERED FROM LIQUID SMOKE

Compound	*Quantity (ppb)*
Pyrene	51,745
Chrysene	48,590
Floranthene	46,260
Benzo[k]fluoranthene	24,710
Benzo[a]pyrene	15,933
Dibenz[a,h]anthracene	14,238
Benzo[g,h,i]perylene	13,058
Benzo[e]pyrene	11,660
Perylene	4,400
Dibenz[a,j]anthracene	3,516

Adapted from Obiedzinski, M., *Acta Aliment. Pol.*, 3, 183, 1977.

had large amounts of PAH present,[25] demonstrating that significant amounts of PAH can be effectively removed from liquid smokes.

However, it is not appropriate to assume that liquid smoke preparations are totally free of PAH. For example, Potthast[26] evaluated 12 commercial liquid smoke products and found from less than 0.05 to 2600 μg/ℓ of benzo[a]pyrene, while White et al.[27] found trace amounts of some three- and four-ring PAH compounds in 3 of 7 commercial liquid smokes they evaluated. In addition, from 25 to 3800 ppb of benzo[a]pyrene was found in the resinous condensates that eventually settled out of these liquid smokes during storage. Therefore, if someone were attempting to be economical and utilize all the material in a liquid smoke container, certain batches of smoke-flavored food could be high in PAH.

Two recent studies have clearly demonstrated that smoke source can dramatically influence both PAH levels and types of compounds present in smoke. In an interesting study evaluating the PAH composition found in various types of commercial briquets,[158] it was found that coal- and wood-base charcoal briquets contained numerous types of PAH ranging up to 13,000 ppb while lava stone and pressed sawdust briquets had no detectable PAH. A portion of these data is summarized in Table 10, and as can be seen, isomeric forms were not positively identified. This study was initiated to determine whether the alleged use of petroleum pitch, coal tar pitch, and high-boiling petroleum distillers as binders in preparing cooking briquets was apparent relative to PAH composition. Three different coal-based briquets had from 36 to 50 individual PAH that ranged from 5400 to 7100 ppb. In the case of three wood-derived briquets, 30 to 53 individual PAH compounds were identified with total concentrations ranging from 2500 to 13,000 ppb. It is of interest to note that benzo[a]pyrene was not found in any samples.

In another study,[159] the PAH composition of mesquite wood smoke and hickory wood smoke were compared. As seen in Table 11, a total of 31 such compounds having a composite composition of 1250 mg/kg of wood were found in the mesquite smoke vs. 22 compounds having a total concentration of 688 mg/kg in the condensed hickory smoke. This is quite interesting in light of the fact of the dramatic growth in certain parts of the U.S. of mesquite-grilled foods. Mesquite is a dense wood that is especially noted for its high smoldering temperature which in turn can significantly encourage PAH formation. The reason for the high thermal degradation temperature associated with mesquite is probably the fact that mesquite was found to be composed of 64% lignin whereas hickory used in the study had only 17.7% lignin[159] and as discussed in an earlier chapter, lignin is the last major wood component to thermally degrade.

IV. PAH IN FOODS

It is not the intention of this section to provide an inclusive listing of every study that has reported upon the presence of PAH in foods. Suffice it to say that literally many hundreds of such studies are in the literature, and most of these data have already been reviewed.[28-49]

At this point it should also be noted that mutagenic effects and PAH content are not unique to smoked foods but are apparently inherent to most heat-processed foods. Method and degree of heat processing, as well as food composition, seem to be important factors.[50-65] The fact that cooking increases PAH levels is clearly evident from the data shown in Table 12.

As can be seen in Table 13, the types of foods in which PAH have been found is as diverse as food itself. If it has been heat processed, it probably contains some amount of PAH. Also, the direct drying of foods with fossil fuels seems to be a major PAH source in products such as malt[66] and certain cereals.[67] For example, Fritz[67] found that when cereals were dried using fuel oil as the heat source, the 3,4-benzopyrene content in the cereals increased 140 to 250% over using natural gas. When lignite briquets were used, amounts increased 220 to 530% and when raw lignite was used, the 3,4-benzopyrene content increased 830 to 900%.

Lawrence and Das[143] measured PAH in various commercial puffed breakfast cereals, and as seen in Table 14, they all had detectable levels with rice being the lowest and wheat the highest. These products are interesting in that they probably have both environmental PAH as well as PAH formed during processing. In the case of puffed rice only one PAH was detected whereas the other cereals had up to four such compounds.

Similar types of data for barley malt were reported by Lawrence and Weber,[65] but as seen in Table 15, the total PAH content among the three commercial sources varied dramatically from 2.2 to 237.9 μg/kg. Undoubtedly the PAH in malt would carry over into the final alcoholic beverage.

The charcoal broiling of meats presents a unique situation whereby heat, smoke, and lipid

Table 10
PAH COMPOSITION (ppb) OF VARIOUS COMMERCIAL CHARCOAL BRIQUETS

	Briquet source					
Compound	**Coal**	**Coal**	**Coal**	**Wood**	**Wood**	**Wood**
Naphthalene	587	102	22	450	735	73
Benzo[b]thiophene	127	—	—	—	—	—
Dimethylbenzofuran	40	—	—	—	—	—
Dimethylbenzofuran	185	—	—	—	—	—
Dimethylbenzofuran	35	—	—	—	—	—
Methylbenzo[b]thiophene	15	—	—	—	—	—
Methylnaphthalene	484	109	52	472	323	82
Methylbenzo[b]thiophene	33	—	—	—	—	—
Methylnaphthalene	405	140	63	360	132	78
Biphenyl	211	91	40	402	161	86
Dimethylbenzo[b]thiophene	21	—	—	—	—	—
Ethylnaphthalene	78	30	18	138	72	49
Ethylnaphthalene	—	—	—	—	—	21
Dimethylnaphthalene	226	69	20	163	132	—
Dimethylnaphthalene	110	226	169	129	465	—
Dimethylnaphthalene	471	83	22	208	232	—
Dimethylnaphthalene	31	—	—	162	—	—
Dimethylnaphthalene	155	—	—	112	—	—
Acenaphthene	41	—	—	—	68	—
Methylbiphenyl	167	79	26	90	228	—
Methylbiphenyl	104	—	—	75	30	—
Dibenzofuran	728	109	140	943	228	174
Methylethylnaphthalene	—	—	—	72	—	—
Trimethylnaphthalene	30	109	11	70	27	—
Trimethylnaphthalene	28	143	85	142	91	—
Trimethylnaphthalene	56	47	81	559	46	—
Trimethylnaphthalene	55	—	184	—	82	—
Trimethylnaphthalene	30	185	714	—	73	—
Trimethylnaphthalene	150	137	—	463	47	—
Trimethylnaphthalene	198	—	—	—	—	—
Fluorene	95	20	52	129	43	62
Trimethylnaphthalene	402	—	—	—	—	—
Dimethylbiphenyl	43	—	—	—	—	—
Methyldibenzofuran	320	263	147	536	169	—
Xanthene	360	90	162	—	338	—
Tetramethylnaphthalene	70	110	—	200	—	—
Isobutylnaphthalene	30	—	—	—	—	—
Methylfluorene	133	50	96	247	38	63
Methylfluorene	55	—	—	370	42	85
Methylfluorene	101	—	—	209	—	—
Tetramethylnaphthalene	30	—	—	279	211	—
Tetramethylnaphthalene	157	—	—	—	—	—
Dibenzothiophene	75	—	—	—	—	—
Phenanthrene	145	141	74	815	338	259
Anthracene	86	40	—	215	—	57
Phenylindole	50	—	—	—	—	31
Phenylnaphthalene	27	—	92	161	56	25
Dimethylphenanthrene	36	25	37	61	38	150
Fluoranthene	20	35	—	257	—	150
Pyrene	23	55	37	160	26	98
Trimethylphenanthrene	29	—	148	—	—	—
Chrysene	30	114	162	139	—	83

Table 10 (continued)
PAH COMPOSITION (ppb) OF VARIOUS COMMERCIAL CHARCOAL BRIQUETS

	Briquet source					
Compound	Coal	Coal	Coal	Wood	Wood	Wood
Tetramethylnaphthalene	—	—	70	—	—	—
Tetramethylnaphthalene	—	81	287	—	—	—
Tetramethylnaphthalene	—	219	—	—	—	—
Methylphenanthrene	—	94	92	279	—	—
Methylphenanthrene	—	89	81	214	—	—
Methylphenanthrene	—	91	81	279	—	—
Methylphenanthrene	—	141	225	190	—	—
Dimethylphenanthrene	—	122	214	110	21	214
Dimethylphenanthrene	—	20	19	—	—	—
Trimethylphenanthrene	—	—	37	621	—	—
Tetramethylphenanthrene	—	164	—	—	—	—
Tetramethylphenanthrene	—	1,675	1,718	113	43	—
Benz[a]anthracene	—	—	96	—	—	—
Perylene	—	51	—	—	—	—
Methyldibenzofuran	—	—	—	643	127	250
Methyldibenzofuran	—	—	—	250	85	53
Methyldibenzofuran	—	—	—	365	—	39
Methyldibenzofuran	—	—	—	—	—	42
Benzo[h]quinoline	—	—	—	—	34	—
Phenylindole	—	—	—	—	—	36
Phenylindole	—	—	—	—	—	35
Methylphenanthizene	—	—	—	—	47	—
Methylphenanthizene	—	—	—	—	42	—
Dimethylphenanthrene	—	—	—	108	34	—
Dimethylphenanthrene	—	—	—	123	—	—
Dimethylphenanthrene	—	—	—	139	—	—
Dimethylphenanthrene	—	—	—	150	—	—
Benzonaphthofuran	—	—	—	150	—	—
Methylfluoranthene	—	—	—	90	—	53
Methylfluoranthene	—	—	—	27	—	—
Methylfluoranthene	—	—	—	135	—	—
Methylfluoranthene	—	—	—	134	—	—
Benzo[c]phenanthrene	—	—	—	—	—	21
Benzo[a]anthracene	—	—	—	—	—	34
Dimethylbenzo[c]-phenanthrene	—	—	—	33	—	—
Total	7,118	5,349	5,574	12,969	5,057	2,509

Adapted from Kushwaha, S. C., Clarkson, S. G., and Mehkeri, K. A., *J. Food Saf.*, 7, 177, 1985.

pyrolysis products can all contribute to the production of PAH. In the mid-1960s reports started to appear which pointed out that this type of food preparation can lead to significant amounts of PAH production. In one of the first studies[50] it was reported that a well-done 1.1-kg steak grilled over charcoal contained 8 μg/kg of benzo[a]pyrene which is equivalent to the amount generated by smoking 600 cigarettes. Other studies[52,58] demonstrated that PAH levels were most closely related to the amount of fat originally present in the grilled food, the closeness of the food to the heat source, and the exposure time of the food to the heat source. In a similar study,[61] it was found that many home and camping food preparation techniques can result in relatively large amounts of PAH. For example, the grilling of

Table 11
PAH COMPOSITION OF MESQUITE AND HICKORY WOOD SMOKE

	μg/kg in	
Compound	**Mesquite**	**Hickory**
Pyrene	155	104
Benzo[a]pyrene	74	41
Benzo[e]pyrene	36	20
1-Methylpyrene	27	18
Dibenzo[a,e]pyrene	4	ND
Dibenzo[a,i]pyrene	3	ND
Dibenzo[a,h]pyrene	3	ND
Dibenzo[a,l]pyrene	2	ND
Indeno[1,2,3-c,d]pyrene	53	24
Fluoranthene	162	94
Benzo[b]fluoranthene	34	22
Benzo[j]fluoranthene	24	11
Benzo[k]fluoranthene	35	14
Benzo[a]fluorene	66	24
Benzo[b]fluorene	31	17
Anthanthrene	22	7
Anthracene	47	31
Benzo[a]anthracene	60	38
2-Methylanthracene	11	4
9-Methylanthracene	3	1
Dibenzo[a,c]anthracene	5	ND
Dibenzo[a,h]anthracene	2	ND
9,10-Diphenylanthracene	3	ND
Phenanthrene	204	114
1-Methylphenanthrene	17	7
2-Methylphenanthrene	21	15
3,6-Dimethylphenanthrene	6	ND
Chrysene + triphenylene	72	49
Picene	7	ND
Perylene	14	4
Benzo[g,h,i]perylene	47	29
Total	1250	688

Note: ND = not detected.

Adapted from Maga, J. A., *J. Agric. Food Chem.*, 34, 249, 1986.

frankfurters in the flames of a log fire produced 212 μg/kg of benzo[a]pyrene whereas when grilling was carried out over the embers, the level dropped to 7.7 μg/kg. As can be seen in Table 16, the method cooking frankfurters dramatically influenced PAH concentration. For example, using the open flames of a log fire produced the highest PAH levels with the PAH concentration dropping by a factor of 10 if cooking was done using just the embers from the same fire. Surprisingly, using pine cones, which contain a large amount of resinous material, as a fuel source was not as detrimental as expected. A small amount of PAH was found if charcoal was used whereas an increase in PAH was not seen with electric oven or pan frying.

Recently, Lawrence and Das[193] determined the PAH content of four commercial brands of charbroiled hamburgers, and as seen in Table 17, the total measured PAH content ranged from 1.51 to 28.1 μg/kg. The actual identities of the hamburger brands were not given, but

Table 12
INFLUENCE OF COOKING ON THE PAH COMPOSITION (ppb) OF FISH

PAH	Raw	Cooked
Anthrace	0.14	1.25
Pyrene	0.77	4.22
Benzo[a]anthracene	0.08	0.29
Benzo[b]fluoranthene	0.36	1.08
Benzo[k]fluoranthene	0.06	0.27
Benzo[a]pyrene	0.05	0.54
Benzo[e]pyrene	0.16	0.89
Benzo[g,h,i]perylene	0.04	0.93
Total	1.66	9.47

Adapted from Obana, H., Hori, S., Tanaka, R., and Kashimoto, T., *J. Food Hyg. Soc. Jpn.*, 25, 35, 1984.

Table 13
FOODS IN WHICH PAH HAVE BEEN IDENTIFIED

Fruits and vegetables	Cooking oils
Citrus	Corn oil
Fruit	Coconut oil
Cabbage	Cottonseed oil
Endive	Olive oil
Kale	Palm oil
Leeks	Palm kernel oil
Lettuce	Peanut oil
Potatoes	Rape oil
Spinach	Soy oil
Tomatoes	Sunflower oil
Dairy products	Meat products
Butter	Bacon
Smoked cheese	Charbroiled meats
Coffee whitener	Frankfurters
Infant milk formula	Ham
Margarine	Sausages
Milk	Smoked meats
Milk powder	Smoked poultry
Cereal products	Fish and seafood products
Bran cereals	Canned shrimp
Puffed cereals	Frozen shrimp
Malt	Lobster
Rice	Lobster paste
Rye	Oysters
Wheat	Scallops
Toast	Shellfish
Legumes	Smoked fish
Cottonseed	Smoked mussels
Peanuts	Smoked oysters
Soybeans	Miscellaneous
Drinks	Copra
Alcoholic beverages	Eggs
Coffee	Olives
Mineral water	Sugar

Table 14
PAH DISTRIBUTION AND CONTENT (μg/kg) IN COMMERCIAL PUFFED CEREALS

PAH	Rice	Corn	Oats	Wheat
F	1.2	1.2	1.9	3.2
B[a]A	ND	ND	0.16	ND
DMBA	ND	ND	ND	ND
B[b]F	ND	0.05	0.08	0.09
B[k]F	ND	0.02	ND	0.04
B[a]P	ND	0.03	ND	0.04
Total	1.2	1.30	2.14	3.37

Note: ND = not detected.

Adapted from Lawrence, J. F. and Das, B. S., *Int. J. Environ. Anal. Chem.*, 24, 113, 1986.

Table 15
PAH COMPOSITION (μg/kg) OF COMMERCIAL BARLEY MALT

	Sample		
PAH	A	B	C
NDMA	T	13	150
FL	0.8	2.6	26
PY	1.1	3.2	48
B[b]FL	0.1	0.1	0.5
DMP	0.1	0.2	6.2
B[a]A	0.1	0.2	4.2
Per	T	0.1	0.4
B[a]P	T	T	0.3
D[a,c]A	T	T	0.4
D[a,h]A	T	T	1.2
Pi	T	T	0.3
IP	T	T	0.4
Total	2.2	16.5	237.9

Note: T = trace.

Adapted from Lawrence, J. F. and Weber, D. F., *J. Agric. Food Chem.*, 32, 794, 1984.

it is apparent from these data that dramatic differences exist. It is not known if the actual time and method of preparation or differences in fat content was the primary factor responsible for these differences, but apparently one of the four companies is aware of the significance of PAH and is controlling their process/product sufficiently to produce a hamburger with low PAH.

In another study, Lawrence and Weber[64] compared PAH in commercially prepared char-broiled vs. fried hamburgers, and as seen in Table 18, the charbroiled products had a greater

Table 16
INFLUENCE OF COOKING TECHNIQUE ON PAH CONCENTRATION (μg/kg) in FRANKFURTERS

	Concentration		
Technique	**Minimum**	**Maximum**	**Average**
Log fire	2.8	171	41.4
Log fire embers	ND	16.3	4.3
Pine cone fire	2.2	21.6	12.9
Charcoal fire	ND	0.8	0.2
Electric oven	ND	ND	ND
Pan frying	ND	ND	ND

Note: ND = not detected.

Adapted from Larsson, B. K., Sahlberg, G. P., Eriksson, A. T., and Busk, L. A., *J. Agric. Food Chem.*, 31, 867, 1983.

Table 17
PAH DISTRIBUTION AND CONTENT (μg/kg) IN COMMERCIAL CHARBROILED HAMBURGERS

	Brand			
PAH	**A**	**B**	**C**	**D**
F	0.73	6.2	15.4	14.7
B[a]A	0.29	2.7	4.9	5.8
DMBA	0.09	1.5	ND	1.7
B[b]F	0.23	1.9	2.1	3.6
B[k]F	0.10	1.2	ND	1.2
D[ah]A	0.07	1.2	ND	1.2
Total	1.51	14.46	23.6	28.1

Note: ND = not detected.

Adapted from Lawrence, J. F. and Das, B. S., *Int. J. Environ. Anal. Chem.*, 24, 113, 1986.

number and concentration of PAH than the fried hamburgers. In light of this, it would be interesting to know if certain of the major hamburger chains who openly advertise that their product is flame broiled are aware of the consequences relative to PAH of this method of preparation.

It should be remembered that PAH can be found in a wide cross section of our food supply because of environmental contamination. A classical example is the presence of such compounds in raw shellfish[64] and the distribution of PAH in wheat[65] where the bran portion was found to contain up to ten times more total PAH than in the endosperm. This should be a factor to consider in the promotion of bran-based breakfast cereals. Data of these type are summarized in Table 19 and 20. Also, any food that undergoes heating during refining and/or processing can be expected to contain PAH. For example, as seen in Table 21, the toasting of bread can result in the formation of increased levels of benzo[a]pyrene.

Table 18
PAH CONTENT (μg/kg) OF COMMERCIALLY CHARBROILED AND FRIED HAMBURGERS

	Charbroiled		Fried	
PAH	A	B	A	B
FL	6.2	5.9	0.1	0.4
PY	14	19	0.4	0.5
B[b]FL	0.5	1.7	T	T
DMP	T	0.6	T	0.1
B[a]A	2.1	2.8	0.1	0.2
Per	0.9	1.1	T	T
B[a]P	1.5	2.9	T	0.1
D[a,c]A	1.0	0.6	T	T
D[a,h]A	2.1	1.3	T	T
Pi	1.0	1.6	T	T
IP	3.6	5.8	T	T
AN	1.1	0.8	T	T
D[a,e]P	0.4	0.7	T	T
D[a,h]P	0.3	0.4	T	T
Total	34.7	45.2	0.6	1.3

Note: T = trace.

Adapted from Lawrence, J. F. and Weber, D. F., *J. Agric. Food Chem.*, 31, 18, 1983.

Table 19
PAH CONTENT OF MILLED WHEAT FRACTIONS AND BREAKFAST CEREALS

Product	PAH content (μg/kg)
Mill fractions	
Whole wheat	9.7
Wheat bran	55.2
Wheat flour	5.1
Breakfast cereals	
Wheat bran	59.5
Puffed wheat	20.3
Whole wheat	18.6
Whole grain oats	6.9
Corn bran	6.7
Flaked milled corn	5.7

Adapted from Lawrence, J. F. and Weber, D. F., *J. Agric. Food Chem.*, 32, 794, 1984.

It should be remembered that nonheat-treated foods can also contain PAH most probably through the effect of environmental contamination, some of which can be smoke discharged from a food smoking operation. For example, the PAH content of lettuce grown at different locations as well as in the open air as compared to a greenhouse situation varied from 4.8 to 94 μg/kg fresh lettuce weight.[185] The lower values were associated with lettuce grown

Table 20
PAH DISTRIBUTION AND CONTENT (μg/kg) IN COMMERCIAL BRAN CEREALS

	Brand					
PAH	A	B	C	D	E	F
F	0.46	1.4	1.5	1.6	1.8	1.3
B[a]A	0.07	ND	0.02	0.06	0.16	0.11
DMBA	0.04	ND	ND	ND	ND	ND
B[b]F	0.06	ND	0.09	0.11	ND	0.63
B[k]F	0.03	ND	0.03	0.05	0.04	0.06
B[a]P	ND	0.02	0.03	0.11	ND	0.07
Total	0.66	1.42	1.67	1.93	2.00	2.17

Note: ND = not detected.

Adapted from Lawrence, J. F. and Das, B. S., *Int. J. Environ. Anal. Chem.*, 24, 113, 1986.

Table 21
THE BENZO[A]PYRENE CONTENT (μg/kg) OF TOASTED BREAD

Product	Benzo[a]pyrene content
Untoasted bread	0.23
Toasted bread — 3 min	0.39
Toasted bread — 5 min	0.56

Adapted from Lintas, C. and DeMatthaeis, M. C., *Food Cosmet. Toxicol.*, 17, 325, 1979.

Table 22
BENZO[A]PYRENE LEVELS IN SOME DUTCH FOODS

Foods	Median (μg/kg)
Potatoes, vegetables, meat, and egg products	0.4
Smoked meat products	0.25
Smoked fish	0.5
Tea leaves	0.8
Tea extract	0.01
Instant coffee	0.09
Endive	1.1
Milk	0.03

Adapted from Vaessen, H. A. M. G., Schuller, P. L., Jekel, A. A., and Wilbers, A. A. M. M., *Toxicol. Environ. Chem.*, 7, 297, 1984.

under glass and the authors concluded that PAH in lettuce resulted primarily from the deposition of PAH-containing particles present in ambient air.

Vegetable oils and fats represent another food group that has usually low PAH levels, but in these foods there is potential concern due to the widespread and relatively high consumption of these products.[186] Usually the major PAH source is combustion-associated

Table 23
ESTIMATED DUTCH MAXIMUM DAILY DIETARY PAH INTAKE (μg)

Compound	Intake
Fluoranthene	10.4
Pyrene	5.1
Chrysene	5.0
Benzo[b,j,k]fluoranthene	1.7
Benzo[g,h,i]perylene	0.9
Benzo[a]pyrene	0.5
Indeno[1,2,3-c,d]pyrene	0.3

Adapted from Vaessen, H. A. M. G., Schuller, P. L., Jekel, A. A., and Wilbers, A. A. M. M., *Toxicol. Environ. Chem.*, 7, 297, 1984.

atmospheric contamination of the raw material source during its growth cycle. In the case of butter for example, contaminated forage or feed is consumed by cattle and the PAH is transmitted to the milk and eventually concentrated in the resulting butter. They[186] also found that crude coconut oil had a PAH concentration of 4600 μg/kg due to the fact that the copra starting material was dried using a heat source that produced smoke. However, a sample of commercially deodorized coconut oil had a PAH level of 92 μg/kg. The authors postulated that the deodorization by steam distillation of this oil source is quite effective in removing PAH-type compounds. When the types of PAH in the two products were compared, it was found that the deodorized oil had a lower concentration of the lower molecular weight PAH due to its distillation during processing.

The Dutch have conducted surveys as to the PAH content of their food supply and as seen in Tables 22 and 23, a wide range of foods were considered with maximum daily PAH amounts from an average diet being 23.9 μg/kg.

V. METHODS OF PAH ANALYSIS

Since certain countries have set maximum limits on the amounts of PAH that can be associated with food, it is imperative that simple, fast, and accurate analytical techniques are available for the routine analysis of foods. Also, these techniques must function over a wide range of food compositional properties (liquids to solids, high to low fat, protein, and carbohydrate).

Historically, procedures were developed that measured only one specific PAH, usually benz[a]pyrene. However, in light of present PAH knowledge, one would have to question this practice, especially in light of the fact that many other PAH compounds may be present in much higher concentrations than this one specific PAH. Also, the carcinogenic properties of other PAH compounds may actually be many times more significant than that of benz[a]pyrene. The general problems with PAH isolation and quantitation have been well reviewed.[68,69]

One general problem with most analytical techniques is the recovery effectiveness of the compound(s) in question over a wide concentration range. In the case of PAH, data as shown in Table 24 are typical. Utilizing high performance liquid chromatography (HPLC) in conjunction with fluorescence detection, Lawrence and Das[193] showed that at relatively low concentrations some PAH have low recovery rates whereas with other PAH, recovery rates are lowered as concentration increases.

Table 24
RECOVERY RATES (%) OF PAH AS INFLUENCED BY CONCENTRATION

	Level added (µg/kg)		
PAH	1.0	5.0	25.0
B[a]A	70.7	50.8	69.8
B[b]F	49.6	57.5	79.0
B[a]P	81.2	92.5	92.3
D[a,i]P	78.6	56.4	20.6

Adapted from Lawrence, J. F. and Das, B. S., *Int. J. Environ. Anal. Chem.*, 24, 113, 1986.

Table 25
RECOVERY RATES (%) OF PAH AT RELATIVELY LOW CONCENTRATIONS

	Level added (µg/kg)		
PAH	1.0	2.0	4.0
FL	119	111	99
PY	106	142	120
B[a]A	103	98	95
DMP	97	96	88
Pi	93	96	91
IP	92	86	86
DPA	89	82	86
D[a,c]A	88	79	71
D[a,c]P	78	74	77
B[b]FL	76	77	83
D[a,i]P	71	82	90
D[a,h]P	70	77	71
D[a,h]A	61	57	60
B[a]P	38	48	50
Per	28	36	44
An	24	34	45

Adapted from Lawrence, J. F. and Weber, D. F., *J. Agric. Food Chem.*, 32, 789, 1984.

Utilizing a similar technique, Lawrence and Weber[64] reported, as seen in Table 25, that some PAH can give recovery rates substantially above 100% even at relatively low concentrations, and thus to accurately quantitate all PAH in each specific food system, one should determine individual PAH recovery rates of the spiked food over the concentration range that is normally expected in that food. Once these specific recovery rates are known, initial data can be factored to obtain a more accurate quantitative picture.

Another potential problem is the fact that since certain PAH may be present in only trace amounts, care must be taken in selecting an analytical technique that is sensitive at low concentrations specifically to the compound(s) in question. As shown in Table 26, the limits

Table 26
PAH DETECTION LIMITS USING HPLC VS. GC

PAH	Detection limit (pg) HPLC	GC
F	18.4	30
B[a]A	13.6	30
DMBA	12.0	30
B[b]F	12.3	30
B[a]P	12.5	60
D[a,h]A	15.6	60
D[a,i]P	264.4	300

Adapted from Lawrence, J. F. and Das, B. S., *Int. J. Environ. Anal. Chem.*, 24, 113, 1986.

of detection are currently lower for HPLC for all PAH evaluated as compared to gas chromatography (GC). This is the primary reason that most prefer HPLC over GC for the composite analysis of PAH.

The problem of potential contamination from solvents and environmental sources during the purification of PAH prior to actual analysis was recently clearly demonstrated by Grimmer and Naujack.[194] At the nanogram range of PAH with which they were working, solvents, glassware, and laboratory air were major sources of contamination. Considering these factors, they were still able to demonstrate the presence of basic nitrogen-containing aromatic compound (N-PAC) which are known to also be carcinogenic in ham.

It can safely be stated that any sensitive and accurate technique currently used for total PAH quantitation is lengthy and tedious. In addition, it usually involves multiple techniques which in turn causes questions as to absolute quantitation due to potential losses during transfer, etc. Some of the general steps usually involved include saponification of the food sample and subsequent solvent extraction, concentration steps involving liquid-liquid partition and column chromatography, separation procedures such as TLC, GC, or HPLC, and final quantitation by ultraviolet, fluorescence, flame ionization, or mass spectrometry. Thus, the total procedure is not only time consuming but usually involves rather expensive equipment.

Howard et al.[70,71] utilized thin-layer chromatography (TLC) to separate and quantitate PAH. Numerous similar procedures have been reported.[72-76] In general, TLC is relatively simple but if a complex mixture of PAH is present, complete separation of all compounds can be somewhat difficult which in turn hinders positive compound identification and quantitation. In addition, care must be taken to prevent PAH photodegradation during separation.

Gas-liquid chromatography (GLC) has also found application for PAH analysis[77-80,89] but difficulty in separating isomeric PAH forms, the need for relatively expensive capillary columns and selective detectors or the need to form derivatives does not make GLC an all inclusive method for PAH analysis. Also, elution times as long as 2 hr are needed for complete elutions.[160]

Most of the more recently published procedures for PAH analysis have utilized HPLC techniques.[64,65,81-88,193,195-197] Although the general method is far from simple, it appears to lend itself to the effective separation and quantitation of a wide variety of PAH compounds.

All of the above techniques have direct application to the analysis of foods but another major analytical concern involves the biological monitoring of humans who may be exposed to PAH in a food smoking operation, for example. Becher and Bjorseth[87] described the extraction of PAH and its metabolites from human urine followed by subsequent HPLC analysis using selective and sensitive fluorescence detection methods.

VI. PAH METABOLISM

Considering the fact that all adults and most children have been exposed to PAH from various dietary and environmental sources, one should ask the question as to metabolically what happens to PAH when ingested. Through much research it has been demonstrated that a series of closely associated enzymes are capable of breaking down PAH in animal cells, but that an alteration in the balance of these enzyme activities can dramatically alter resulting types and amounts of PAH metabolites.[90]

The key enzyme appears to be aryl hydrocarbon hydroxylase[91] which is actually a complex of enzymes[92] in which cytochrome P-450 plays a major role. By this enzymatic action PAH can be converted to phenol and numerous dihydrodihydroxy derivatives, which in turn can either enzymatically or chemically be converted to conjugated sulfates, glucuronides, and mercapturic acids. Another enzyme, epoxide hydrase, is important in converting PAH epoxides to dihydrodiols.[93] In addition, PAH that have methyl groups as ring components are important metabolically since they yield hydroxy-methyl derivatives.[94] Thus, most if not all humans have the general capability to metabolically handle PAH, but the unanswered questions deal with how much and for how long can the human effectively metabolize PAH before they can cause significant metabolic harm.

The scientific literature has also caused some confusion due to the fact that the development of arylhydroxylase activity is apparently dependent upon the location of administering PAH.[40] For example, if given intratracheally, PAH is eliminated in less than a day whereas if it is administered subcutaneously, it is still detectable 6 months later. It has also been shown[40] that in experimental animals, higher resulting tumor rates are associated with subcutaneous injection as compared to intratracheal administration. In addition, the dose/effect response was more negative with subcutaneous injection since 500 times the quantity needed to produce skin cancer was required to produce cancer in the respiratory tract. From these data some would conclude that PAH ingestion is not as dangerous as epicutaneous or subcutaneous application, but this is a narrow and potentially misleading assumption. If enough PAH is present in the diet, there is no doubt that the risk for cancer is increased.

The concentration of PAH in certain workplace environments can exceed by factors of more than 1000 the PAH concentration in normal atmosphere.[87] This should be a cause for health concern but some argue from a metabolic standpoint that this may not sound as bad as it is since only a small portion of PAH found in the environment is bioavailable due to the fact that PAH is adsorbed to particles. Also, some argue that man may be less sensitive than experimental animals with respect to PAH-induced cancer.

From a metabolic standpoint there is potential concern that certain of the PAH metabolically oxidized products are potentially more of a health concern than nonoxidized PAH. Also to be considered is the fact that hydroxylated PAH may induce the development of carcinomas before there is the opportunity for these compounds to be detoxified via activated glucuronic acid from the body.[40]

Sound scientific metabolic study, especially utilizing humans, is difficult in determining the potential interaction of both carcinogenic and noncarcinogenic PAH compounds with other toxic and nontoxic compounds. For example, certain noncarcinogenic PAH have a cocarcinogenic effect when combined with certain benzopyrenes whereas other PAH have been reported to inhibit carcinogenesis.[40]

It has also been reported[40] that the metabolic resorbability of certain PAH is improved by the consumption of caffeine. The obvious question arising from this observation is does a heavy coffee or tea drinker who also consumes foods high in PAH have a higher cancer risk than a person who does not consume caffeine-containing drinks? Other potential PAH/diet constituent interactions probably exist but research in this area is still lacking.

Sullivan[187] discussed the potential for free radicals to serve as intermediates in the metabolism of PAH. For example, with benzo[a]pyrene, at least four different types of free radicals have been observed as influenced by reaction conditions. The simplest free radical form involves a one-electron reduction or a one-electron oxidation resulting in the formation of an anion or cation radical. Also, another free radical species can form when benzo[a]pyrene is heated above its melting point. In addition, numerous oxygenated free radicals such as 6-oxybenzo[a]pyrene have been identified.[187] From a biological standpoint, clear evidence for biological involvement of the anion form has been proposed. From a smoke generation standpoint, the melting of benzo[a]pyrene is especially interesting. Cavalieri and Rogan[199] demonstrated that the one-electron oxidation radical form can cause PAH carcinogenesis by the direct application of PAH to rat mammary gland.

Cation radicals have been shown[187] to react with certain nucleophiles at position 6. Also, the benzo[a]pyrene cation radical is quenched by numerous nitrogenous compounds and antioxidants. In addition, this radical has been shown[187] to result in adduct formation with DNA. Oxy radicals also bind strongly with DNA and compounds such as caffeine.

VII. PAH STABILITY

Primarily from a pollution concern, research has been conducted to determine the stability of PAH once they are emitted into the ecosystem. For example, Thomas et al.[95] demonstrated that PAH are closely associated with soot particles and as such do not appear to be present in air in their free form. However, Korfmacher et al.[96] have reported that once absorbed onto coal fly ash, PAH vary in their susceptibility to photodecompose. They found that benzo[a]pyrene, pyrene, and anthracene all were efficiently photolyzed in liquid solution but were highly resistant to photodegradation when absorbed onto ash. In contrast, phenanthrene and fluoranthene were resistant to photodecomposition both in liquid and ash systems. The relationship of these observations to food systems is not known, and thus, would serve as an interesting study to determine the stability of various PAH in food systems as influenced by such factors as solids content, packaging, and exposure to light with time.

Jager[97] reported that PAH emitted into the atmosphere are quickly degraded by several pathways including photooxidation, reaction with atmospheric oxidants, and reactions with other chemical pollutants such as nitrogen oxides and nitric acid.

Cimberle et al.[99] have also noted that photodecomposition of PAH is an important pathway for their decomposition, with time of year relative to the amount and intensity of sunshine being important factors to consider. Thus, during cloudy days their decomposition may be quite low thereby making them more of an environmental concern.

As pointed out by Seifert,[98] PAH decomposition can also be a potential problem during their analysis. He found a 50% reduction in benzo[a]pyrene after a few minutes of separation on TLC. Thus, if analytical precautions are not taken, there is the potential to grossly underestimate PAH levels in food systems. The stability of other PAH compounds during analysis should also be investigated.

In contrast, Gunther and Westlake[198] reported that when certain PAH were applied at a rate of 10 ppm to the surfaces of mature oranges still on the tree, a persistance of up to 175 days resulted indicating that at least on oranges, PAH is not completely photooxidized in short periods of time. A portion of their data is summarized in Table 27.

PAH contamination of certain ecosystems has resulted in the detection of PAH in raw shellfish such as mussels and oysters.[100-103] Thus, if these products are smoked and found to contain PAH, one cannot be sure of the portion, if any, of PAH which results from the

Table 27
DEGRADATION AND PERSISTANCE OF PAH (10 ppm) WHEN APPLIED TO ORANGES

	Half-life (days)	
PAH	Degradation	Persistence
B[a]P	1	175
B[ai]P	1.5	12
B[a,h]A	2	125
An	1.5	175

Adapted from Gunther, F. A. and Westlake, W. E., *Residue Rev.*, 17, 81, 1967.

Table 28
PAH AMOUNTS IN CONTAMINATED OYSTERS

Compound	Exposure time (days)	Depuration time (days)	Concentration in oyster (μg/g)
Anthracene	2	0	5.6
	2	7	1.2
	8	7	0.4
	8	23	0.1
Fluoranthene	2	0	5.0
	2	7	1.7
	8	7	1.4
	8	23	0.4
Benz[a]anthracene	2	0	2.8
	2	7	1.9
	8	7	1.0
	8	23	0.3
Benzo[a]pyrene	2	0	0.36
	2	7	0.40
	8	7	0.20
	8	23	0.12

Adapted from Lee, R. F., Gardner, W. S., Anderson, J. W., Blaylock, J. W., and Barwell-Clarke, J., *Environ. Sci. Technol.*, 12, 832, 1978.

smoking process. Apparently, PAH are relatively stable in live shellfish since in a study reported by Lee et al., oysters that were intentionally exposed to anthracene, fluoranthene, benz[a]anthracene, and benzo[a]pyrene for up to 8 days and then subjected for up to 23 days of depuration or purging still had measurable PAH levels. These data are summarized in Table 28.

VIII. NITROSAMINES

Nitrosamines are a class of compounds that can be formed by the action of nitrite on secondary and tertiary amines, and are powerful biological agents many of which have been shown to be carcinogenic.[104] They have been identified in a wide variety of foods[105-135] but most research has centered around their presence in cured and smoked meat products since the intentional addition of nitrate/nitrite and spices to these products is a common practice.[136-138] However, as pointed out by White,[139] cured meat products only account for approximately one sixth of the nitrate consumed in the U.S., while vegetables account for over four fifths. Also, two thirds of the nitrite that reaches the gut comes from saliva while cured meats account for the other third.

To make matters more interesting relative to nitrosamine formation in smoked foods, Walker et al.[140] demonstrated that phenols can actually catalyze nitrosamine formation in model systems. Thus, it can be assumed that in the presence of nitrite, phenols present in wood smoke can form nitrosamines. They also demonstrated that this same reaction can lead to high results during certain nitrosamine analyses.

In light of the above information, the utilization of liquid smoke preparations in a brine that contains nitrite has to be seriously questioned since nitro- and nitrosophenols can form under these conditions.

In a more recent study, Mandagere et al.[141] actually reacted wood smoke condensate with combinations of nitrite and cysteamine, cysteine, or cystine. Small amounts of certain nitrosamines were found in all systems, but the combination of wood smoke, nitrite, and cysteamine produced significant amounts of nitrosamines.

Kalve and Kann[142] also reported upon the formation of various nitrogen oxides from wood smoke that could enter into nitrosamine formation. Smoke generation temperature was the primary factor governing the amounts of nitrogen oxides formed with the higher the temperature, the greater the oxide levels.

Recently, the influence of phenol and 2,6-dimethoxyphenol (syringol), two common wood smoke components, on the rates of *N*-nitrosomorpholine, *N*-nitrosopyarolidine, and *N*-nitrosopyrrolidine production has shown that syringol significantly increased the production of all three whereas phenol had no influence on *N*-nitrosomorpholine levels but dramatically increased *N*-nitrosopyrrolidine production by approximately 350%. Thus, it would appear that smoke high in certain phenols may have either beneficial or negative effects relative to nitrosoamine formation.

Two recently discovered nitroso compounds are *N*-nitrosothiazolidine and *N*-nitrosothiazolidine-4-carboxylic acid[163,171-179] which have been most closely associated with cured, smoked foods such as cured meats and bacon. The reasons these two specific compounds are of potential concern are that *N*-nitrosothiazolidine has been indicated as being genotoxic in microbial and hepatocellular test systems[180] while *N*-nitrosothiazolidine-4-carboxylic acid has been shown to induce diabetes in experimental animals.[175] This latter compound thus has been implicated as the compound responsible for the high levels of juvenile diabetes associated with mothers who consumed large amounts of smoked meats during pregnancy.[175] Sen et al.[179] reported that fried bacon that had been commercially directly smoked had levels of *N*-nitrosothiazolidine-4-carboxylic acid of up to 13,700 ppb when fried.

The levels of nitrosamines in commercial bacon have been measured[181-184] and in the most recent study[184] the average amounts in milligrams per kilogram of *N*-nitrosodimethylamine in the rasher and fried-out fat were 4.0 and 5.3; 0.7 and 1.5 for *N*-nitrosopiperidine; 17 and 32 for *N*-nitrosopyrrolidine; and 8.9 and 3.4 for *N*-nitrosothiazolidine. From these data they concluded that the average daily intake of *N*-nitrosopyrrolidine from fried bacon was 76 ng per person. Also, they found that many consumers of bacon use the fried-out fat for other cooking purposes which would further elevate their intake of nitrosamines.

The reader is directed to an excellent series of papers[164-170] that cover nitrosamine formation, analysis, and safety issues in smoked and heated foods.

IX. TOXICOLOGICAL PROPERTIES OF WOOD SMOKE

Several Polish groups have reported on the toxicological properties of commercial liquid smokes.[143-145] Based on standard procedures, Kozlowski et al.[143] determined that the LD_{50} doses for their liquid smoke were 2750 mg/kg of mouse body weight, 3750 mg/kg of rat body weight, 192 mg/kg of cat body weight, and greater than 5000 mg/kg of pig body weight. Histopathological analysis of organs revealed no difference from control groups not fed liquid smoke.

Fitko et al.[144] fed Polish large white swine daily doses of 200, 600, 1000, or 1400 mg/kg of body weight of liquid smoke while Swiss albino mice and Wistar rats received diets containing 0.1, 0.5, 1.0, and 2.0% liquid smoke for 3 months. In the case of pigs, pathological changes in the digestive tract, liver, kidneys, and lungs were apparent but biochemical and hematological blood changes were not observed. In both pigs and rodents, increasing amounts of liquid smoke in the diet decreased body weight gains even though feed consumption increased thus indicating that liquid smoke adversely influenced digestion and/or absorption.

Fitko et al.[145] also determined LD_{50} values in mice and rats administered liquid smoke. The LD_{50} for female mice was 2420 mg/kg while for male mice it was 2700 mg/kg of body weight. In the case of female rats, an LD_{50} of 3550 mg/kg was calculated while for male rats it was 4010 mg/kg. In the animals that died, gross pathological changes were observed including hyperemia and hemorrhages in the stomach and duodenal mucosa, and hyperemia and edema of the kidneys and liver. In test animals that did not die, the higher doses of liquid smoke caused parenchymatous and fatty degeneration of liver and kidneys and acute to subacute catarrhal inflammation of the stomach and small intestine mucosa. The authors concluded that liquid smoke demonstrated low acute toxicity and the physiological changes associated with high dosages were typical to those of similar compounds. However, because of the potential wide differences in composition and concentration of liquid smoke preparations, these studies should not be taken as average LD_{50} values for smoke in general.

X. MUTAGENIC PROPERTIES OF WOOD SMOKE

Most scientists now believe that compounds that can be shown to possess bacterial mutagenic activity have a potential cause for concern in our diet. However, the question as to whether all compounds that are demonstrated to be mutagenic are also carcinogenic is still being debated. In any event, a few studies have appeared reporting on the mutagenic properties of wood smoke although more studies have appeared on the mutagenic properties of individual phenols,[146,150] which can be the major components of wood smoke. For the most part, individual phenols have given negative results relative to mutagenic activity.

One relatively recent study has reported upon the mutagenic properties of phenolic smoke fractions and their isolated phenols.[151] They found no mutagenic activity, at five concentrations up to 5000 μg per plate, associated with phenol, cresols, 2,4-dimethylphenol, benzcatechine, syringol, eugenol, vanillin, and guaiacol. However, when smokehouse smoke was condensed and fractionated, certain phenolic fractions did give an increase in the number of revertants. The authors postulated that apparently there can be small amounts of mutagenic impurities in wood smoke that give a positive test for mutagen formation. At this point the specific compounds responsible are not known. However, as pointed out by Haugen and Peak,[152] complex mixtures of aromatic compounds can actually suppress mutagenic activity, thus possibly resulting in erroneous conclusions. Since so little work relative to smoke has been reported in this area, more well-designed experiments are required in order to resolve this very important issue.

Obviously it would be difficult to obtain direct evidence positively correlating the consumption of smoked foods to specific forms of cancer, but general data, such as the fact that stomach and colon cancer occurrences are higher in countries or populations that have significant ingestion levels of PAH,[153] strongly point to the fact that since most smoked foods contain various types and amounts of PAH, these foods can serve as one source as precursors for cancer.

In actual animal studies utilizing mice and rats, Kozlowski et al.[143] reported on the influence of smoke extract consumption on inducing carcinomas and teratogenic effects. They calculated that a human population would consume approximately 0.10 mg of smoke extract per 1 kg of body weight while in their animal studies consumption was in the range of 16 to 21 mg of extract per kilogram body weight. No signs of neoplastic or mutagenic activity were noted in the test animals.

REFERENCES

1. **Simon, S., Rypinski, A. A., Tauber, F. W., Pencyla, R. M., and Westerberg, D. O.,** Effect of cellulose casing on the absorption of polycyclic hydrocarbons in wood smoke by absorbents, *J. Agric. Food Chem.*, 17, 1128, 1969.
2. **Rusz, J., Kopalova, M., and Pelikanova, K.,** Effect of some factors on 3,4-benzpyrene content of smoke for smoking food, *Prum. Potravin*, 22, 106, 1971.
3. **Lenges, J.,** Smoking of meat products, *Rev. Ferment. Ind. Aliment.*, 27, 53, 1972.
4. **Toth, L. and Blaas, W.,** The effect of smoking technology on the content of carcinogenic hydrocarbons in smoked meat products. II. Effect of smoldering temperature of wood and of cooling, washing and filtering of smoke, *Fleischwirtschaft*, 52, 1419, 1972.
5. **Toth, L. and Blaas, W.,** The effect of smoking technology on the content of carcinogenic hydrocarbons in smoked meat products. I. Effect of various smoking methods, *Fleischwirtschaft*, 52, 1121, 1972.
6. **Kersken, H.,** Smoking today, *Fleischwirtschaft*, 54, 1721, 1974.
7. **Potthast, K.,** Problems of smoking meat and meat products, *Fleischwirtschaft*, 55, 1492, 1975.
8. **Prihavec, D. and Polic, M.,** Basic characteristics of manufacture of a smoke concentrate; the properties of this concentrate, *Tehnol. Mesa*, 17, 349, 1976.
9. **Steinig, J.,** Benzo[a]pyrene contents of smoked fish in relation to smoking procedure, *Z. Lebensm. Unter. Forsch.*, 162, 235, 1976.
10. **Lenges, J., Vo Thi, N. B., and Declerck, D.,** Effect of smoking technology on the contamination of smoked fish with 3,4-benzopyrene, *Rev. Ferment. Ind. Aliment.*, 31, 97, 1976.
11. **Steinig, J. and Schreiber, W.,** Minimization of the 3,4-benzopyrene content of smoked fish products by modification of the smoking process, *Inf. Fleischwirtschaft*, 24, 126, 1977.
12. **Potthast, K., Eigner, G., and Eichner, R.,** Reduction of content of polycyclic hydrocarbons in smoked meat products, *Ber. Landwirtscha.*, 55, 817, 1977.
13. **Hamm, R.,** Analysis of smoke and food, *Fleischwirtschaft*, 57, 92, 1977.
14. **Borys, A., Klossowska, B., Obiedzinski, M., and Olkiewicz, M.,** Influence of combustion conditions on the composition of carbonylic, carboxylic, polynuclear hydrocarbons, and phenolic compounds present in wood smoke, *Acta Aliment. Pol.*, 3, 335, 1977.
15. **Potthast, K.,** Smoking procedures and their effect on 3,4-benzpyrene and other smoke components in smoked meat products, *Fleischwirtschaft*, 58, 371, 1978.
16. **Hardy, R. and McGill, A. S.,** Smoking of foods: methods and some toxicological aspects, *Process Biochem.*, 14(11), 2, 1979.
17. **Potthast, K.,** The influence of smoking technology on the composition of polycyclic hydrocarbons in smoked meat products, smoke condensates and in waste gases from smoking plants, *Fleischwirtschaft*, 59, 1515, 1979.
18. **Raja, K. C. M., Mahendraker, N. S., and Morrjani, M. N.,** Benzpyrene in smoked poultry, *J. Food Sci. Technol. India*, 17, 156, 1980.
19. **Potthast, K.,** Dark smoking at high smokehouse temperature, *Fleischwirtschaft*, 61, 1630, 1981.
20. **Dikun, P. P., Kostenko, L. D., Shendrikova, L. D., Emshanova, A. V., and Nikitin, B. N.,** Contents of 3,4-benzpyrene in fish processed by various smoking technologies, *Rybn. Khoz.*, 5, 78, 1981.
21. **Sirami, J.,** Application of smoking processes, *Viandes Produits Carnes*, 2(6), 8, 1981.
22. **Rhee, K. S. and Bratzler, L. J.,** Polycyclic hydrocarbon composition of wood smoke, *J. Food Sci.*, 33, 626, 1968.
23. **Obiedzinski, M. and Borys, A.,** Identification of polynuclear aromatic hydrocarbons in wood smoke, *Acta Aliment. Pol.*, 3, 169, 1977.

24. **Colmsjo, A. L., Zebuhr, Y. U., and Ostman, C. E.,** Polycyclic aromatic compounds in curing smoke, *Z. Lebensm. Unters. Forsch.*, 179, 308, 1984.
25. **Obiedzinski, M.,** Determination of polycyclic aromatic hydrocarbons in wood smoke, *Acta Aliment. Pol.*, 3, 183, 1977.
26. **Potthast, K.,** The composition of smoking preparations, *Fleischwirtschaft,* 54, 183, 1974.
27. **White, R. H., Howard, J. W., and Barnes, C. J.,** Determination of polycyclic aromatic hydrocarbons in liquid smoke flavors, *J. Agric. Food Chem.*, 19, 143, 1971.
28. **Toth, L.,** Carcinogenic substances in smoked meat products, *Fleischwirtschaft,* 49, 1611, 1969.
29. **Tilgner, D. J.,** Carcinogenic hydrocarbons in food, *Gordian,* 71, 2, 1971.
30. **Luks, D. and Lenges, J.,** Some analytical and practical aspects of the contamination of smoked meat products by carcinogenic compounds, *Rev. Ferment. Ind. Aliment.*, 28, 111, 1973.
31. **Hamm, R.,** Polycyclic carcinogenic compounds in meat products and toxic trace elements in meat, *Med. Weter.*, 29, 708, 1973.
32. **Filipovic, J. and Karakas, R.,** The impact of smoking and the content of carcinogens in meat products, *Hrana Ishrana,* 14, 447, 1973.
33. **Gilbert, J. and Knowles, M. E.,** The chemistry of smoked foods: a review, *J. Food Technol.*, 10, 245, 1975.
34. **Henning, W.,** How dangerous are smoke-flavored seasonings?, *Ernaehrungswirtschaft,* 3, 108, 1976.
35. **Polic, M.,** The chemistry of smoke and smoke concentrates: the effect and importance of individual components for the meat products, *Tehnol. Mesa,* 17, 345, 1976.
36. **Fritz, W. and Soos, K.,** Studies on the content of carcinogenic hydrocarbons in smoked foods in the German Democratic Republic and in Hungary, *Nahrung,* 21, 951, 1977.
37. **Walker, E. A.,** Some facts and legislation concerning polycyclic aromatic hydrocarbons in smoked foods, *Pure Appl. Chem.*, 49, 1673, 1977.
38. **Bories, G., Tchimbakala, A., and Tchimbakala, E.,** A survey of food contamination by 3,4-benzpyrene, *Ann. Nutr. Aliment.*, 32, 811, 1978.
39. **Potthast, K.,** The occurrence of 3,4-benzpyrene in smoked meat products, seasonings and other foods, *Fleischerei,* 29(5), 67, 1978.
40. **Potthast, K.,** The problem of 3,4-benzopyrene in smoked meat products, *Fleischwirtschaft,* 58, 38, 1978.
41. **Mossanda, K., Poncelet, F., Fouassin, A., and Mercier, M.,** Detection of mutagenic polycyclic aromatic hydrocarbons in African smoked fish, *Food Cosmet. Toxicol.*, 17, 141, 1979.
42. **Binnemann, P. H.,** Benso[a]pyrene contents of meat products, *Z. Lebensm. Unters. Forsch.*, 169, 447, 1979.
43. **Schober, B.,** Smoke components and their effects, *Lebensm. Ind.*, 26, 262, 1979.
44. **Mossanda, K.,** Polycyclic aromatic hydrocarbons in smoked fish from central Africa, *Rev. Ferment. Ind. Aliment.*, 35, 132, 1980.
45. **Potthast, K.,** Recent figures for benzo[a]pyrene contents in meat products, *Fleischwirtschaft,* 60, 1941, 1980.
46. **Nistor, C., Raileanu, L., Pacular, T., Ichim, I., Mancas, D., Stoleru, F., Sinpetru, L., and Lamasanu, F.,** Contamination of smoked foods with polycyclic aromatic hydrocarbons, *Igiena,* 29, 345, 1980.
47. **Larsson, B. K.,** Polycyclic aromatic hydrocarbons in smoked fish, *Z. Lebensm. Unters. Forsch.*, 174, 101, 1982.
48. **Ruchkovskii, B. S., Tikhtin, L. A., and Palyura, A. N.,** Contents of benzo[a]pyrene in sausage products, *Vopr. Pitan.*, 1, 62, 1982.
49. **Afolabi, O. A., Adesulu, E. A., and Oke, O. L.,** Polynuclear aromatic hydrocarbons in some Nigerian preserved freshwater fish species, *J. Agric. Food Chem.*, 31, 1083, 1983.
50. **Lijinsky, W. and Shubik, P.,** Benzo[a]pyrene and other polynuclear hydrocarbons in charcoal-broiled meat, *Science,* 145, 53, 1964.
51. **Lijinsky, W. and Shubik, P.,** Polynuclear hydrocarbon carcinogens in cooked meat and smoked food, *Ind. Med. Surg.*, 34, 152, 1965.
52. **Lijinsky, W. and Ross, A. E.,** Production of carcinogenic polynuclear hydrocarbons in the cooking of food, *Food Cosmet. Toxicol.*, 5, 343, 1967.
53. **Tilgner, D. J. and Daun, H.,** Polycyclic aromatic hydrocarbons in smoked food, *Residue Rev.*, 27, 19, 1969.
54. **Caurie, M., Lee, T. C., Salomon, M., and Chichester, C. O.,** Hot smoke fish curing, *J. Natl. Sci. Counc. Sri Lanka,* 2, 77, 1974.
55. **Engst, R. and Fritz, W.,** Food-hygenic toxicological evaluation of the occurrence of carcinogenic hydrocarbons in smoked products, *Acta Aliment. Pol.*, 3, 255, 1977.
56. **Yamazaki, H., Minami, J., Ichiakwa, T., and Kondo, M.,** Analysis of polycyclic aromatic hydrocarbons in raw and broiled fishes, *J. Food Hyg. Soc. Jpn.*, 18, 368, 1977.
57. **Lintas, C. and DeMatthaeis, M. C.,** Determination of benzo[a]pyrene in smoked, cooked and toasted food products, *Food Cosmet. Toxicol.*, 17, 325, 1979.

58. **Doremire, M. E., Harmon, G. E., and Pratt, D. E.,** 3,4-benzopyrene in charcoal grilled meats, *J. Food Sci.*, 44, 622, 1979.
59. **Heddeghem, A., Giel, W., and Moor, H.,** Polycyclic aromatic hydrocarbons in smoked and grilled foods, *Meded. Fac. Landbouwwet. Rijksuniv. Gent*, 44, 1221, 1979.
60. **Newburg, D. S. and Concon, J. M.,** Malonaldehyde concentrations in food are affected by cooking conditions, *J. Food Sci.*, 45, 1681, 1980.
61. **Larsson, B. K., Sahlberg, G. P., Eriksson, A. T., and Busk, L. A.,** Polycyclic aromatic hydrocarbons in grilled foods, *J. Agric. Food Chem.*, 31, 867, 1983.
62. **Fretheim, K.,** Polycyclic aromatic hydrocarbons in grilled meat products — a review, *Food Chem.*, 10, 129, 1983.
63. **Bjeldanes, L. F., Morris, M. M., Timourian, H., and Hatch, F. T.,** Effects of meat composition and cooking conditions on mutagen formation in fried ground beef, *J. Agric. Food Chem.*, 31, 18, 1983.
64. **Lawrence, J. F. and Weber, D. F.,** Determination of polycyclic aromatic hydrocarbons in some Canadian commercial fish, shellfish, and meat products by liquid chromatography with confirmation by capillary gas chromatography-mass spectrometry, *J. Agric. Food Chem.*, 32, 789, 1984.
65. **Lawrence, J. F. and Weber, D. F.,** Determination of polycyclic aromatic hydrocarbons in Canadian samples of processed vegetable and dairy products by liquid chromatography with fluorescence detection, *J. Agric. Food Chem.*, 32, 794, 1984.
66. **Ondrisekova, M. and Sedova, V.,** Study of phenolic compounds in smoked malt ordered by Japanese customers, *Kvasny Prum.*, 27(3), 58, 1981.
67. **Fritz, W.,** Formation of carcinogenic hydrocarbons during thermal treatment of foods. VI. Investigation of the contamination of cereals resulting from smoke drying, *Nahrung*, 18, 83, 1974.
68. **Haenni, E. O.,** Analytical control of polycyclic aromatic hydrocarbons in food and food additives, *Residue Rev.*, 24, 41, 1968.
69. **Grimmer, G.,** Analysis of polycyclic aromatic hydrocarbons, *Landwirtsch. Forsch.*, 35, 683, 1979.
70. **Howard, J. W., Teague, R. T., White, R. H., and Fry, B. E.,** Extraction and estimation of polycyclic aromatic hydrocarbons in smoked foods. I. General method, *J. Assoc. Off. Anal. Chem.*, 49, 595, 1966.
71. **Howard, J. W., White, R. T., Fry, B. E., and Turicchi, E. W.,** Extraction and estimation of polycyclic aromatic hydrocarbons in smoked foods. IV. Benzo[a]pyrene, *J. Assoc. Off. Anal. Chem.*, 49, 611, 1966.
72. **Rijk, M. A. H. and Battum, D.,** A rapid method for determining benz[a]pyrene in smoke aroma, *Dtsch. Lebensm. Rundsch.*, 69, 75, 1973.
73. **Kedzierski, B. and Kubacki, S. J.,** A modification of the Howard method of benzo[a]pyrene determination in foods, *Acta Aliment. Pol.*, 3, 175, 1977.
74. **Shiraishi, Y. and Shirotori, T.,** Determination of polycyclic aromatic hydrocarbons in foods. VII. Separation and determination of benzo[a]pyrene and benzo[k]fluoranthene, *J. Food Hyg. Soc. Jpn.*, 20, 345, 1979.
75. **Mossanda, K.,** Polycyclic aromatic hydrocarbons in smoked fish from central Africa. I. Development of a qualitative and quantitative analytical method, *Rev. Ferment. Ind. Aliment.*, 35, 99, 1980.
76. **Segredos, A. N., Sinha-Roy, D., and Jan, E.,** A contribution to the determination of polycyclic aromatic hydrocarbons in smoked foodstuffs via caffeine complexes, *Dtsch. Lebensm. Rundsch.*, 77, 33, 1981.
77. **Grimmer, G. and Bohnke, H.,** Polycyclic aromatic hydrocarbon profile analysis of high-protein foods, oils and fats by gas chromatography, *J. Assoc. Off. Anal. Chem.*, 58, 725, 1975.
78. **Potthast, K.,** Polycyclic aromatic hydrocarbons in smoked meat products. An application of a new method, *Acta Aliment. Pol.*, 3, 195, 1977.
79. **Radecki, A., Lamparczyk, H., Grzybowski, J., and Halkiewicz, J.,** Separation of polycyclic aromatic hydrocarbons and determination of benzo[a]pyrene in liquid smoke preparations, *J. Chromatogr.*, 150, 527, 1978.
80. **Tanner, R. L. and Fajer, R.,** Determination of nitro-polynuclear aromatics in ambient aerosol samples, *Int. J. Environ. Anal. Chem.*, 14, 231, 1983.
81. **Guerrero, H., Biehl, E. R., and Kenner, C. T.,** High-pressure liquid chromatography of benzo[a]pyrene and benzo[ghi]perylene in oil-contaminated shellfish, *J. Assoc. Off. Anal. Chem.*, 59, 989, 1976.
82. **Toussaint, G. and Walker, E. A.,** Use of high-performance liquid chromatography as a clean-up procedure in analysis of polycylic aromatic hydrocarbons in alcoholic beverages, *J. Chromatogr.*, 171, 448, 1979.
83. **Crosby, N. T., Hung, D. C., Philp, L. A., and Patel, I.,** Determination of polynuclear aromatic hydrocarbons in food, water and smoke using high-performance liquid chromatography, *Analyst*, 106, 135, 1981.
84. **Gertz, C.,** Improved method for the quantitative separation of 3,4-benzpyrene(benzo[a]pyrene) from food, *Z. Lebensm. Unters. Forsch.*, 173, 208, 1981.
85. **Joe, F. L., Salemme, J., and Fazio, T.,** High performance liquid chromatography with fluorescence and ultraviolet detection of polynuclear aromatic hydrocarbons in barley malt, *J. Assoc. Off. Anal. Chem.*, 65, 1395, 1982.

86. **Nielsen, T.**, Isolation of polycyclic aromatic hydrocarbons and nitro derivatives in complex mixtures by liquid chromatography, *Anal. Chem.*, 55, 286, 1983.
87. **Becher, G. and Bjorseth, A.**, Determination of exposure to polycyclic aromatic hydrocarbons by analysis of human urine, *Cancer Lett.*, 17, 301, 1983.
88. **Balya, D. R. and Danchik, R. S.**, Chromatographic techniques for characterization of coal tar pitch volatiles, *Am. Ind. Hyg. Assoc. J.*, 45, 260, 1984.
89. **Radecki, A., Lamparczyk, H., Halkiewicz, J., and Grzybowski, J.**, Studies on determination of benzo[a]pyrene in smoke flavorings, *Acta Aliment. Pol.*, 3, 209, 1977.
90. **Brookes, P.**, Mutagenicity of polycyclic aromatic hydrocarbons, *Mutation Res.*, 39, 257, 1977.
91. **Gelboin, H. V., Wiebel, F. J., and Kinoshita, N.**, Aryl hydrocarbon hydroxylase regulation and role in polycyclic hydrocarbon action, *Chemical Carcinogenesis, Part A*, Marcel Dekker, New York, 1974, 309.
92. **Huberman, E., Yamasaki, H., and Sachs, L.**, Independent regulation of two types of aryl hydrocarbons benzo[a]pyrene hydroxylase in mammalian cells, *Int. J. Cancer*, 18, 76, 1976.
93. **Oesch, F.**, Mammalian epoxide hydrases: inducible enzymes cytolysing the inactivation of carbinogenic and cytotoxic metabolites derived from aromatic and definite compounds, *Xenobiotica*, 3, 305, 1973.
94. **Sims, P.**, The metabolism of 7- and 12-methylbenz[a]anthracene and their derivatives, *Biochem. J.*, 105, 591, 1967.
95. **Thomas, J. F., Mukai, M., and Tebbens, B. D.**, Fate of airborne benzo[a]pyrene, *Environ. Sci. Technol.*, 2, 33, 1968.
96. **Korfmacher, W. A., Wehry, E. L., Mamantov, G., and Natusch, D. F. S.**, Resistance to photochemical degradation of polycyclic aromatic hydrocarbons vapor-absorbed on coal fly ash, *Environ. Sci. Technol.*, 14, 1094, 1980.
97. **Jager, J.**, Detection and characterization of nitro derivatives of some polycyclic aromatic hydrocarbons by fluorescence quenching after thin-layer chromatography: application to air pollution analysis, *J. Chromatogr.*, 152, 575, 1978.
98. **Seifert, B.**, Stability of benzo[a]pyrene on silica gel plates for high-performance thin-layer chromatography, *J. Chromatogr.*, 131, 417, 1977.
99. **Cimberle, M. R., Bottino, P., and Valerio, F.**, Decomposition of benzo[a]pyrene deposited on glass fiber filters and exposed to sunlight, *Chemosphere*, 12, 317, 1983.
100. **Iosifidou, H. G., Kilikidis, S. D., and Kamarianos, A. P.**, Analysis for polycyclic aromatic hydrocarbons in mussels *(Myticus galloprovincialis)* from the Thermaikos Gulf, Greece, *Bull. Environ. Contam. Toxicol.*, 28, 535, 1982.
101. **Xoaa, S., Picard-Berube, M., and Gouygou, J. P.**, Polynuclear aromatic hydrocarbons in mussels from the estuary and northwestern Gulf of St. Lawrence, Canada, *Bull. Environ. Contam. Toxicol.*, 31, 41, 1983.
102. **Mix, M. C. and Schaffer, R. L.**, Concentrations of unsubstituted polynuclear aromatic hydrocarbons in bay mussels *(Mytilus edulis)* from Oregon, U.S.A., *Marine Environ. Res.*, 9, 193, 1983.
103. **Lee, R. F., Gardner, W. S., Anderson, J. W., Blaylock, J. W., and Barwell-Clarke, J.**, Fate of polycyclic aromatic hydrocarbons in controlled ecosystem enclosures, *Environ. Sci. Technol.*, 12, 832, 1978.
104. **Magee, P. N.**, Toxicity of nitrosamines: their possible human health hazards, *Food Cosmet. Toxicol.*, 9, 207, 1971.
105. **Ceh, L. and Ender, F.**, A sensitive method for the colorimetric determination of volatile nitrosamines in food products and air, *Food Cosmet. Toxicol.*, 16, 117, 1978.
106. **Dikun, P. P.**, Group evaluation of volatile and nonvolatile N-nitroso compounds in foodstuffs, *IARC (Int. Agency Res. Cancer) Sci. Publ.*, 14, 57, 1976.
107. **Dikun, P. P., Romanova, L. E., Shendrikova, I. A., Reim, N. D., and Yunusova, Z. Y.**, Contents of *N*-nitrosamines in fish and some types of fish preserves, *Rybn. Khoz.*, 8, 69, 1980.
108. **Nieper, L. and Etzel, V.**, Nitrate content in smoked fish muscle, *Arch. Lebensmittelhyg.*, 34, 149, 1983.
109. **Sen, N. P. and Dalpe, C.**, A simple thin-layer chromatographic technique for the semi-quantitative determination of volatile nitrosamines in alcoholic beverages, *Analyst*, 97, 216, 1972.
110. **Spiegelhalder, B., Eisenbrand, G., and Preussmann, R.**, Contamination of beer with trace quantities of *N*-nitrosodimethylamine, *Food Cosmet. Toxicol.*, 17, 29, 1979.
111. **Walker, E. A., Castegnaro, M., Garren, L., Toussaint, G., and Kowalski, B.**, Intake of volatile nitrosamines from consumption of alcohols, *J. Natl. Cancer Inst.*, 63, 947, 1979.
112. **Slack, P. T. and Wainwright, T.**, Hordenine as the precursor of NDMA in malt, *J. Inst. Brew.*, 87, 259, 1981.
113. **Hotchkiss, J. H., Havery, D. C., and Fazio, T.**, Rapid method for estimation of *N*-nitrosodimethylamine in malt beverages, *J. Assoc. Off. Anal. Chem.*, 64, 929, 1981.
114. **Lance, D. G., Barrett, P. A., Kavanagh, T. E., and Clarke, B. J.**, Methods for the determination of volatile nitrosamines in malt, wort and beer, *J. Inst. Brew.*, 87, 19, 1981.

115. **Havery, D. C., Hotchkiss, J. H., and Fazio, T.,** Nitrosamines in malt and malt beverages, *J. Food Sci.*, 46, 501, 1981.
116. **Tauts, O. V. and Kaive, R. E.,** Causes of *N*-nitrosodimethylamine formation in beer, *Tallinna Polutehnilise Inst. Toim.*, No. 537, 38, 1982.
117. **Huxel, E. T., Scanlan, R. A., and Libbey, L. M.,** Formation of *N*-nitrosopyrrolidine from pyrrolidine ring containing compounds at elevated temperatures, *J. Agric. Food Chem.*, 22, 698, 1974.
118. **Telling, G. M., Bryce, T. A., and Althorpe, J.,** Use of vacuum distillation and gas chromatography-mass spectrometry for determination of low levels of volatile nitrosamines in meat products, *J. Agric. Food Chem.*, 19, 937, 1971.
119. **Sen, N. P.,** The evidence for the presence of dimethylnitrosamine in meat products, *Food Cosmet. Toxicol.*, 10, 219, 1972.
120. **Crosby, N. T., Foreman, J. K., Palframan, J. F., and Sawyer, R.,** Estimation of steam-volatile *N*-nitrosamines in foods at the 1 μg/kg level, *Nature*, 238, 342, 1972.
121. **Wasserman, A. E., Fiddler, W., Doerr, R. C., Osman, S. F., and Dooley, C. J.,** Dimethylnitrosamine in frankfurters, *Food Cosmet. Toxicol.*, 10, 681, 1972.
122. **Panalaks, T., Lyengar, J. R., and Sen, N. P.,** Nitrate, nitrite, and dimethyl-nitrosamine in cured meat products, *J. Assoc. Off. Anal. Chem.*, 56, 621, 1973.
123. **Bills, D. D., Hildrum, K. I., Scanlan, R. A., and Libbey, L. M.,** Potential precursors of *N*-nitrosopyrrolidine in bacon and other fried foods, *J. Agric. Food Chem.*, 21, 876, 1973.
124. **Gough, T. A., Goodhead, K., and Walters, C. L.,** Distribution of some volatile nitrosamines in cooked bacon, *J. Sci. Food Agric.*, 27, 181, 1976.
125. **Patterson, R. L. S., Taylor, A. A., Mottram, D. S., and Gough, T. A.,** Localized occurrence of *N*-nitrosopyrrolidine in fried bacon, *J. Sci. Food Agric.*, 27, 257, 1976.
126. **Sen, N. P., Seaman, S., and Miles, W. F.,** Volatile nitrosamines in various cured meat products: effect of cooking and recent trends, *J. Agric. Food Chem.*, 27, 1354, 1979.
127. Nitrite Safety Council, A survey of nitrosamines in sausages and dry-cured meat products, *Food Technol.*, 45, 1980.
128. **Hotchkiss, J. H., Libbey, L. M., and Scanlan, R. A.,** Confirmation of low μg/kg amounts of volatile *N*-nitrosamines in foods by low resolution mass spectrometry, *J. Assoc. Off. Anal. Chem.*, 63, 74, 1980.
129. **Hauser, E., Gruber, E., and Mihalik, K.,** Nitrosamine formation in bacon. Effect of formaldehyde, *Proc. Eur. Meet. Meat Res. Workers*, 27, Vol. II, 480, 1981.
130. **Greenfield, E. L., Vasco, G. A., Legette, L., and Malanoski, A.,** Screening procedure for detection of volatile *N*-nitrosamines in cooked bacon by one-trap mineral oil vacuum distillation and thermal energy analyzer, *J. Assoc. Off. Anal. Chem.*, 65, 1316, 1982.
131. **Greenfield, E. L., Smith, W. J., and Malanoski, A. J.,** Mineral oil vacuum distillation method for nitrosamines in fried bacon, with thermal energy analyzer: collaborative study, *J. Assoc. Off. Anal. Chem.*, 65, 1319, 1982.
132. **Kimoto, W. I., Pensabene, J. W., and Fiddler, W.,** Isolation and identification of *N*-nitrosothiazolidine in fried bacon, *J. Agric. Food Chem.*, 30, 757, 1982.
133. **Pensabene, J. W., Miller, A. J., Greenfield, E. L., and Fiddler, W.,** Rapid dry column method for determination of *N*-nitrosopyrrolidine in fried bacon, *J. Assoc. Off. Anal. Chem.*, 65, 151, 1982.
134. **Havery, D. C., Kline, D. A., Miletta, E. M., Joe, F. L., and Fazio, T.,** Survey of food products for volatile *N*-nitrosamines, *J. Assoc. Off. Anal. Chem.*, 59, 540, 1976.
135. **Sen, N. P. and Seaman, S.,** Volatile *N*-nitrosamines in dried foods, *J. Assoc. Off. Anal. Chem.*, 64, 1238, 1981.
136. **Sen, N. P., Miles, W. F., Donaldson, B., Panalaks, T., and Lyengar, J. R.,** Formation of nitrosamines in a meat curing mixture, *Nature*, 245, 104, 1973.
137. **Sen, N. P., Donaldson, B., Charbonneau, C., and Miles, W. F.,** Effect of additives on the formation of nitrosamines in meat curing mixtures containing spices and nitrite, *J. Agric. Food Chem.*, 22, 1125, 1974.
138. **Gough, T. A. and Goodhead, K.,** Occurrence of volatile nitrosamines in spice premixes, *J. Sci. Food Agric.*, 26, 1473, 1975.
139. **White, J. W.,** Relative significance of dietary sources of nitrate and nitrite, *J. Agric. Food Chem.*, 23, 886, 1975.
140. **Walker, E. A., Pignatelli, B., and Friesen, M.,** The role of phenols in catalysis of nitrosamine formation, *J. Sci. Food Agric.*, 33, 81, 1982.
141. **Mandagere, A. K., Gray, J. I., Skrypec, D. J., Booren, A. M., and Pearson, A. M.,** Role of woodsmoke in *N*-nitrosothiazolidine formation in bacon, *J. Food Sci.*, 49, 658, 1984.
142. **Kalve, R. E. and Kann, Y. M.,** Formation of nitrogen oxides during production of curing smoke, *Tallinna Polutehnilise Inst. Toim.*, No. 489, 15, 1980.
143. **Kozlowski, Z. P.,** Investigations on the chemical, toxicological and biological properties of the Polish smoke extract, *Proc. Eur. Meet. Meat Res. Workers*, 15, 516, 1969.

144. **Fitko, R., Markiewicz, K., Markiewicz, Z., Szuperski, T., and Krzymowska, H.,** The zootechnical indexes and pathological changes in animals after smoke flavouring application in subacute toxicity test, *Acta Aliment. Pol.*, 3, 269, 1977.
145. **Fitko, R., Grabowski, J., Krzymowska, H., Markiewicz, K., and Szuperski, T.,** The acute toxicity of smoke flavouring, *Acta Aliment. Pol.*, 3, 275, 1977.
146. **Florin, F., Rutberg, L., Curvall, M., and Enzall, C. R.,** Screening of tobacco smoke constituents for mutagenicity using the Ames test, *Toxicology*, 15, 219, 1980.
147. **Gilbert, P., Rondelet, J., Poncelet, F., and Mercier, M.,** Mutagenicity of *p*-nitrosophenol, *Food Cosmet. Toxicol.*, 18, 523, 1980.
148. **Gocke, E., Kind, M. T., Eckhardt, K., and Wild, D.,** Mutagenicity of cosmetics ingredients licensed by the European communities, *Mutation Res.*, 90, 91, 1981.
149. **McCann, J., Choi, E., Yamasaki, E., and Ames, B. N.,** Detection of carcinogens as mutagens in the salmonella microsome test: assay of 300 chemicals, *Proc. Natl. Acad. Sci. U.S.A.*, 72, 5135, 1975.
150. **Shahin, M. M., Bugaut, A., Gilard, P., and Kalopissis, G.,** Studies on the mutagenicity of resorcinol and hydroxy-3-(*p*-amino) anilino-6 *N*[*p*-aminophenol] benzoquinone-monoimine-1,4 in *Salmonella typhimurium*, *Mutation Res.*, 78, 213, 1980.
151. **Pool, B. L. and Lin, P. Z.,** Mutagenicity testing in the *Salmonella typhimurium* assay of phenolic compounds and phenolic fractions obtained from smokehouse smoke condensates, *Food Chem. Toxicol.*, 20, 383, 1982.
152. **Haugen, D. A. and Peak, M. J.,** Mixtures of polycyclic aromatic compounds inhibit mutagenesis in the salmonella/microsome assay by inhibition of metabolic activation, *Mutation Res.*, 116, 257, 1983.
153. **Toth, L. and Potthast, K.,** Chemical aspects of the smoking of meat and meat products, *Adv. Food Res.*, 29, 87, 1984.
154. **Colmsjo, A. L., Zebuhr, Y. U., and Ostman, C. E.,** Polycyclic aromatic compounds in curing smoke, *Z. Lebensm. Unters. Forsch.*, 179, 308, 1984.
155. **Muller, W. D.,** Filling and smoking frankfurter-type sausages, *Fleischwirtschaft*, 65, 287, 1985.
156. **Wigand, W. and Jahr, D.,** Benzo[a]pyrene content of smoked meat products, *Fleischwirtschaft*, 65, 980, 1985.
157. **Uherova, R., Dubravicky, J., Smirnov, V., and Hrivnak, J.,** Hygienic aspects of liquid smoke, *Proc. Eur. Meet. Meat Res. Workers*, 31, 516, 1985.
158. **Kushwaha, S. C., Clarkson, S. G., and Mehkeri, K. A.,** Polycyclic aromatic hydrocarbons in barbecue briquets, *J. Food Saf.*, 7, 177, 1985.
159. **Maga, J. A.,** Polycyclic aromatic hydrocarbon (PAH) composition of mesquite *(Prosopis fuliflora)* smoke and grilled beef, *J. Agric. Food Chem.*, 34, 249, 1986.
160. **Vaessen, H. A. M. G., Schuller, P. L., Jekel, A. A., and Wilbers, A. A. M. M.,** Polycyclic aromatic hydrocarbons in selected foods; analysis and occurrence, *Toxicol. Environ. Chem.*, 7, 297, 1984.
161. **Gilbert, J., Knowles, M. E., and McWeeny, D. J.,** Formation of C- and S-nitroso compounds and their further reactions, *J. Sci. Food Agric.*, 26, 1785, 1975.
162. **Virk, M. S. and Issenberg, P.,** Nitrosation of phenol and 2,6-dimethyoxyphenol and its effect on nitrosamine formation, *J. Agric. Food Chem.*, 33, 1082, 1985.
163. **Massey, R. C., Crews, C., Dennis, M. J., McWeeney, D., Startin, J. R., and Knowles, M. E.,** Identification of a major new involatile *N*-nitroso compound in smoked bacon, *Anal. Chim. Acta*, 174, 327, 1985.
164. **Hotchkiss, J. H. and Vecchio, A. J.,** Nitrosamines in fried-out bacon fat and its use as a cooking oil, *Food Technol.*, 38(1), 67, 1985.
165. **Skrypec, D. J., Gray, J. I., Mandagere, A. K., Booren, A. M., Pearson, A. M., and Cuppett, S. L.,** Effect of bacon composition and processing on *N*-nitrosamine formation, *Food Technol.*, 38(1), 74, 1985.
166. **Havery, D. C. and Fazio, T.,** Human exposure to nitrosamines from foods, *Food Technol.*, 38(1), 80, 1985.
167. **Sen, N. P., Seaman, S. W., and Baddoo, P. A.,** *N*-nitrosothiazolidine and nonvolative *N*-nitroso compounds in foods, *Food Technol.*, 38(1), 84, 1985.
168. **Wagner, D. A. and Tannenbaum, S. R.,** In-vivo formation of *N*-nitroso compounds, *Food Technol.*, 38(1), 89, 1985.
169. **Pensabene, J. W. and Fiddler, W.,** Formation and inhibition of *N*-nitrosothiazolidine in bacon, *Food Technol.*, 38(1), 91, 1985.
170. **Scanlan, R. A. and Reyes, F. G.,** An update on analytical techniques for *N*-nitrosamines, *Food Technol.*, 38(1), 95, 1985.
171. **Kimoto, W. I., Pensabene, J. W., and Fiddler, W.,** The isolation and identification of *N*-nitrosothiazolidine in fried bacon, *J. Agric. Food Chem.*, 30, 757, 1982.
172. **Pensabene, J. W. and Fiddler, W.,** *N*-nitrosothiazolidine in cured meat products, *J. Food Sci.*, 48, 1870, 1983.

173. **Sen, N. P., Seaman, S. W., and Baddoo, P.,** *N*-nitrosothiazolidine and nonvolatile *N*-nitroso compounds in foods, *Food Technol.*, 38(1), 84, 1985.
174. **Mandagere, A. K., Gray, J. I., Skrypec, D. J., Booren, A. M., and Pearson, A. M.,** Role of woodsmoke in *N*-nitrosothiazolidine formation in bacon, *J. Food Sci.*, 49, 658, 1984.
175. **Helgason, T., Ewen, S. W. P., Jaffray, B., Stowers, J. M., Outram, J. R., and Pollock, J. R. A.,** *N*-nitrosamines in smoked meats and their relation to diabetes, *IARC (Int. Agency Res. Cancer) Sci. Publ.*, 57, 911, 1984.
176. **Pensabene, J. W. and Fiddler, W.,** Factors affecting the *N*-nitrosothiazolidine content of bacon, *J. Food Sci.*, 48, 1452, 1983.
177. **Pensabene, J. W. and Fiddler, W.,** Formation and inhibition of *N*-nitrosothiazolidine in bacon, *Food Technol.*, 38(1), 91, 1985.
178. **Tricker, A. R., Perkins, M. J., Massey, R. C., Bishop, L., Key, P. E., and McWeeny, D. J.,** Incidence of some non-volatile *N*-nitroso compounds in cured meats, *Food Additives Contaminants*, 1, 245, 1984.
179. **Sen, N. P., Baddoo, P. A., and Seaman, S. W.,** *N*-nitrosothiazolidine and *N*-nitrosothiazolidine-4-carboxylic acid in smoked meats and fish, *J. Food Sci.*, 51, 821, 1986.
180. **Loury, D. J., Byard, J. L., and Shibamoto, T.,** Genotoxicity of *N*-nitrosothiazolidine in microbial and hepatocellular test systems, *Food Chem. Toxicol.*, 22, 1013, 1984.
181. National Research Council, *The Health Effects of Nitrate, Nitrite and N-nitroso Compounds*, National Academy Press, Washington, D.C., 1981.
182. **McCutcheon, J. W.,** Nitrosamines in bacon: a case study of balancing risks, *Public Health Rep.*, 99, 360, 1984.
183. **Havery, D. C. and Fazio, T.,** Human exposure to nitrosamines from foods, *Food Technol.*, 39(1), 80, 1985.
184. **Vecchio, A. J., Hotchkiss, J. H., and Bisogni, C. A.,** *N*-nitrosamine ingestion from consumer-cooked bacon, *J. Food Sci.*, 51, 754, 1986.
185. **Wickstrom, K., Pyysalo, H., Plaami-Heikkila, S., and Tuominen, J.,** Polycyclic aromatic compounds (PAC) in leaf lettuce, *Z. Lebensm. Unters. Forsch.*, 183, 182, 1986.
186. **Hopia, A., Pyysalo, H., and Wickstrom, K.,** Margarines, butter and vegetable oils as sources of polycyclic aromatic hydrocarbons, *J. Am. Oil Chem. Soc.*, 63, 889, 1986.
187. **Sullivan, P. D.,** Free radicals of benzo[a]pyrene and derivatives, *Environ. Health Perspect.*, 64, 283, 1985.
188. **Obana, H., Hori, S., Tanaka, R., and Kashimoto, T.,** Dietary intakes of polycyclic aromatic hydrocarbons, *J. Food Hyg. Soc. Jpn.*, 25, 35, 1984.
189. **Yamazaki, H., Minami, J., Ichikawa, T., and Kondo, M.,** Analysis of polycyclic aromatic hydrocarbons in raw and broiled fishes, *J. Food Hyg. Soc. Jpn.*, 18, 368, 1977.
190. **Badger, G., Kimber, R., and Spotswood, T.,** Polycyclic aromatic hydrocarbons for the pyrolysis of various organic compounds, *Nature*, 187, 663, 1960.
191. **Schmeltz, I. and Hoffmann, D.,** *Carcinogenesis*, Vol. 1, Frendenttal, R. I. and Jones, P. W., Eds., Raven Press, New York, 1976.
192. **Fedonin, V. F., Kuznetsov, V. V., Bershova, T. M., and Basmanov, P. I.,** Content of benz[a]pyrene in dispersed smoking medium in relation to aerosol particle size, *Myasn. Ind. SSSR*, 12, 35, 1978.
193. **Lawrence, J. F. and Das, B. S.,** Determination of nanogram/kilogram levels of polycyclic aromatic hydrocarbons in foods by HPLC with fluorescence detection, *Int. J. Environ. Anal. Chem.*, 24, 113, 1986.
194. **Grimmer, G. and Naujack, K. W.,** Gas chromatographic profile analysis of basic nitrogen-containing aromatic compounds (azaarenes) in high protein foods, *J. Assoc. Off. Anal. Chem.*, 69, 537, 1986.
195. **Arashidani, K., Yoshikawa, M., and Kodama, Y.,** A simplified analysis of polynuclear aromatic hydrocarbons in soil and sediment by high performance liquid chromatography, *Eisei Kagaku*, 31, 24, 1985.
196. **Tsuji, I., Koide, K., Moriyama, H., Tanabe, K., and Matsushita, H.,** Analysis of polynuclear aromatic hydrocarbons in shellfishes, seabottom sediments and sea water by high performance liquid chromatography, *J. Food Hyg. Soc. Jpn.*, 26, 50, 1985.
197. **Friocourt, M. P., Bodennec, G., and Berthou, F.,** Determination of polyaromatic hydrocarbons in scallops *(Pecten maximus)* by UV fluorescence and HPLC combined with UV and fluorescence detectors, *Bull. Environ. Contam. Toxicol.*, 34, 228, 1985.
198. **Gunther, F. A. and Westlake, W. E.,** Residue behavior of polynuclear hydrocarbons on and in oranges, *Residue Rev.*, 17, 81, 1967.
199. **Cavalieri, E. and Rogan, E.,** Role of radical cations in aromatic hydrocarbon carcinogenesis, *Environ. Health Perspect.*, 64, 69, 1985.

Chapter 11

TOBACCO SMOKE

I. INTRODUCTION

This monograph deals with smoke and foods, and thus with a little latitude, one could consider tobacco and its smoke as a food source since tobacco is pyrolyzed and a portion of the resulting smoke is intentionally inhaled. As a result, much research has been conducted on tobacco and its smoke chemistry and it has been found to be a complex and fascinating area. As a result, certain portions of tobacco smoke research have direct application in understanding smoked food chemistry. Other obvious by-products of tobacco smoke research include health and environmental issues associated with smoke in general.

The number of compounds identified in tobacco products and their resulting smokes number literally into the thousands and thus it would be almost impossible to compile a complete listing in the framework of this review. In addition, it has been predicted that the actual number of compounds that are probably present approaches 100,000, thus making tobacco and its resulting smoke one of the most diverse and complex chemical systems known to man.

These compounds can be derived from a number of sources including the tobacco itself, the environment in which it was grown, the portion of the plant utilized, the method of curing and processing, and the temperature and amount of oxygen present during the actual smoking process. Thus, it is not surprising that a large number of compounds, many of which have not yet been identified, and their contribution to flavor and health established, are present in tobacco and tobacco smoke systems.

Also to be considered are the various forms in which tobacco can be utilized. In the case of cigarette, pipe, and cigar tobaccos, the generation and inhalation of smoke are the major objectives whereas with chewing tobacco, smoke obviously is not a factor, but the enzymatic degradable and water soluble constituents, along with certain aroma properties are of major concern.

Because of degree of consumption, the vast majority of research has centered around the cigarette since it has been implicated with certain health issues, but recently the role of chewing tobacco and resulting health has also triggered major research projects in an attempt to better understand tobacco chemistry and physiology.

II. THE TOBACCO COMBUSTION PROCESS

Obviously the tobacco combustion process is somewhat dependent on the form of tobacco usage which in turn will influence the resulting smoke composition. For the purpose of this review, the combustion process associated with the commercially available machine-rolled cigarette will be reviewed.

Essentially there are four separate components to combustion in a filtered cigarette. These include the combustion, pyrolysis, and distillation zones, along with the filter itself. The combustion zone is the red, glowing portion of a lit cigarette. In this area combustion is almost complete, thus resulting in the conversion of tobacco to essentially carbon dioxide, carbon monoxide, and water. During puffing, combustion occurs primarily at the outside edges of the cigarette, thus also degrading the paper wrapper, while between draws combustion is most pronounced in the central portion of the cigarette. Thus, during the active inhalation of cigarette smoke, paper, as well as tobacco, is contributing to smoke composition.

Tobacco formulation and cigarette construction can also influence combustion temperature with a high combustion temperature resulting in a faster burning cigarette thus permitting fewer puffs per cigarette. In addition, tar and PAH formation are influenced by combustion temperature with higher combustion temperatures producing more by-products. Combustion temperature can be reduced through the addition of additives[1] but if the resulting cigarette is not puffed frequently, it may extinguish itself.

The pyrolysis zone is just behind the combustion zone and between draws has a temperature of approximately 600°C, while during puffing the pyrolysis temperature increases to 900 to 1000°C.[1] At approximately 600°C tobacco is kindled and starts to burn, while at 450°C charring takes place. However, even at a temperature of 300°C, vapors from tobacco can condense and form smoke. Because of the temperature of pyrolysis and a reducing gas rich in hydrogen, carbon dioxide is converted to carbon monoxide and a wide variety of hydrocarbons are produced. The hydrocarbons in turn condense on small nuclei emitted from the combustion zone producing a dense smoke cloud composed of spherical particles having a fairly uniform size.[2] The density of the resulting smoke in turn can be controlled by permitting air incorporation through minute perforations in the wrapper.

The next component is the distillation zone, which obviously becomes shorter as the cigarette is burned. In this zone the smoke temperature drops rapidly to approximately 40°C. Because of the resulting cooling effect, smoke constituents having a high boiling point tend to condense and as a result these compounds can be thermally activated or reacted again as the cigarette shortens and the pyrolysis zone moves closer to the filter. Apparently, little if any mechanical filtration occurs in this zone. However, because of the porous wrapper, smoke and its components can diffuse out of the cigarette while the constituents to be inhaled can be further diluted by room air drawn through the wrapper.

The last potential reaction site in most cigarettes is the filter whose primary purpose is to physically and/or chemically trap particulate matter and certain toxic components that may still be present in the gaseous phase. Dependent upon design and the strength of puffing, most filters have an efficiency in the range of 20 to 75%. Most of the more efficient filters have an air vent incorporated into their design which serves to not only dilute smoke but also to slow the passage of smoke through the filter which in turn permits more particle deposition in the filter. However, previously deposited volatile compounds can be reincorporated into the smoke phase during intensive puffing. Also, a filter can be designed to be too effective relative to the deposition of compounds having intermediate volatility, since apparently these types of compounds contribute significantly to flavor. Thus, if they are effectively removed via the filter, resulting flavor suffers.

The material used in filter construction can also influence the types and amounts of compounds removed. For example, a carbon filter is quite effective in removing irritants such as acrolein, while cellulose acetate filters are especially effective in removing polar compounds such as phenols, and filters composed of hydrocarbon polymers are best for the selective removal of nonpolar compounds. Regardless of filter composition, they are in general equally efficient in removing particulate smoke components, nicotine, and tars.

III. TOBACCO FORMS

All common tobacco comes from numerous variants of the species *Nicotiana tabacum* but the composition of tobacco can vary significantly dependent upon numerous factors such as variety, environment, location within the plant, and the method of curing.[4] These factors can result in a 3-fold difference in sterols, a 5-fold variation in phenols, and up to a 30-fold difference in alkaloids. Environmental conditions such as the type of soil, availability and type of minerals, organic matter, and water, as well as the amount of light and temperature during growth are important considerations. If there is little nitrogen available, acetate from the tricarboxylic acid cycle will be low in tobacco since it will have been used to produce carbohydrates, lipids, oils, resins, and polyterpines. Low nitrogen will also result in low nicotine but high gum levels.[3]

Various portions of the tobacco plant can find their way into different tobacco products. For example, leaves and stems near the base of the plant, called scrape trash or trash lugs, are used for granulated smoking tobacco, while leaves higher up the plant are used for cigarettes, and tips are used for chewing tobacco. As one goes further up the plant, carbohydrate content decreases but the nicotine and nitrogen content increase and the degree of combustibility decreases, thus altering resulting smoke composition. A tobacco that has high combustibility has relatively low amounts of total particulates, cresols, phenols, hydrogen cyanide, and benzanthracene but larger amounts of acetaldehyde, acrolein, and carbon monoxide.

Tobacco curing usually begins with leaf dehydration which results in the destruction of chlorophylls as well as the hydrolysis of starch to simple sugars, the breakdown of proteins to amino acids, and the decomposition of pectins into pectic and uronic acids, and methanol.

Some tobacco is flue cured whereby leaves are exposed to a humid environment and the temperature is gradually raised to 75°C over a 5- to 7-day period. The resulting product is then redried and aged for periods of up to several years during which slow chemical changes continue to occur. Flue-cured tobacco also tends to accumulate phenols and other smoke constituents from the hardwood fires used as a heat source in the drying process. Snuff is usually exclusively made from this type of tobacco.

Tobacco being used for the production of cigarettes is first usually air cured where some hydrolysis and oxidation of sugars to acids occurs. Some of the phenols also oxidize and are polymerized. After storage in the dry state, the product is permitted to undergo several fermentation cycles at 45°C and 60% relative humidity. Naturally occurring oxidative and hydrolytic enzymes further decrease sugar content. Also, the levels of organic and amino acids, pectins, hemicelluloses, gums resins, and volatile oils are reduced.

Most cigarettes are produced from blends of various tobaccos each of which contributes a characteristic property to the resulting flavor and smoke due to basic differences in their chemical composition. For example, nitrogen compounds contribute strength, the tannins and resins give aroma, sugars give mildness, and cell-membrane components contribute sharpness to odor. A common tobacco blend includes varying proportions of Virginian, which is usually flue cured, Burley, Maryland, and Oriental. The Virginian is used to impart good flavor and aroma due to its relatively high sugar content. Burley is usually the major tobacco in the blend and has a low carbohydrate content, thus burning well and producing a mild smoke. Also, Burley has the ability to absorb flavoring agents and humectants which are normally added to cigarettes. Maryland tobacco burns well and has moderate amounts of nitrogenous compounds and acids. Oriental tobaccos are usually added for their contribution to aroma.

Currently, most cigarettes also have a certain proportion of reconstituted tobacco sheet added. This product is manufactured using traditional paper-making techniques with the continuous sheet that is formed being chopped and blended with natural tobacco. Due to alterations in its physical structure brought about by using air or carbon dioxide to puff or expand the product, its combustion rate is faster than nontreated tobacco. Also, the process permits partial reduction of naturally occurring tars and nicotine.

Nontobacco components that can be added to cigarettes include casing solutions composed of sugars, syrups, honey, liquorice, chocolate, or balsams, humectants such as glycerol and glycols, and alcohol-soluble flavor compounds like menthol, peppermint, cloves, and cinnamon. Many of these compounds add to the complexity of resulting tobacco smoke.

Most paper used to form cigarettes is made from flax and is usually treated with various chemicals to modify its rate of burning so as to burn at the same rate and intensity as the tobacco portion of a cigarette.[3]

Table 1
MAJOR COMPOSITION OF CIGARETTE SMOKE

Component	Mass %
Nitrogen	59.0
Carbon dioxide	13.6
Oxygen	13.4
Particulate matter	8.2
Carbon monoxide	3.2
Argon	1.0
Methane	0.3
Hydrogen	0.1
Water vapor	1.2
Hydrocarbons	0.5
Carbonyls	0.4
Hydrogen cyanide	0.1
Other gases	0.2

Adapted from Shephard, R. J., *The Risks of Passive Smoking*, Oxford University Press, New York, 1982.

Tobacco used in cigars is air dried and then fermented, thus it contains less protein, alkaloids, and natural sugars than cigarette tobacco. Also, a cigar is wrapped in a tobacco leaf rather than paper. The levels of phenols and aromatic hydrocarbons are similar in cigars and cigarettes.[5,6] However, the condensates from cigars contain more PAH and pyridine[7] and a higher concentration of carbon monoxide than cigarettes since most cigars are tightly rolled and thus burn less readily.[8] It has also been reported that the carbon monoxide associated with cigar smoke is unprotonated and, therefore, is readily absorbed by the body.[9]

Most pipe tobacco can have significant amounts of nontobacco additives such as liquorice to bind the tobacco together, sugar, humectants, and various flavoring agents. Also, a large amount of the humectants and their resulting thermal decomposition by-products can be present in the inhaled smoke. Combustion of added sugars also increases the amounts of furfural, nicotine, and tar present during inhalation. It has been reported that the pipe smoke contains more compounds per volume of smoke than most cigars or cigarettes.[7]

IV. MAJOR TOBACCO SMOKE CONSTITUENTS

As with all smoke, tobacco smoke has both a particulate and gaseous phase; however, many compounds can be found in both phases,[10] and as can be seen in Table 1, gases make up by far the vast majority of the mass associated with cigarette smoke.

As would be expected, many compounds associated with tobacco smoke are known to present certain health risks and the U.S. government has classified compounds into three general categories relative to the degree of risk. Some of these data are summarized in Table 2.

Carbon monoxide levels in cigarettes are of major concern and have been found to range from approximately 5 to 21 mg per cigarette.[3] Its concentration in inhaled cigarette smoke varies with the length of the burning cigarette with the first puffs producing relatively small amounts but as the cigarette gets shorter, less air is present and thus, carbon monoxide concentration increases. Also, carbon monoxide concentration increases as puff length, and thus smoke volume, increases.

Table 2
COMPOUNDS POSSESSING VARYING DEGREES OF RISK IN TOBACCO SMOKE

Most likely hazards
  Carbon monoxide, nicotine, tar
Probable health hazards
  Acrolein, cresol, hydrogen cyanide, nitric oxide, nitrogen dioxide, phenol
Possible health hazards
  Acetaldehyde, acetone, acetonitrile, acrylonitrile, ammonia, benzene, 2,3-butadione, butylamine, carbon dioxide, crotononitrile, dimethylamine, endrin, ethylamine, formaldehyde, furfural, hydrogen sulfide, hydroquinone, methacrolein, methanol, methylamine, nickel compounds, pyridine

Adapted from Shephard, R. J., *The Risks of Passive Smoking*, Oxford University Press, New York, 1982.

Table 3
PAH IN CIGARETTE SMOKE

Compound	Smoke concentration (μg/cigarette)
Pyrene	50—200
Methylpyrene	50—300
Fluoranthrene	100—260
Benzo[g,h,i]perylene	60
Benzo[a]pyrene	10—50
Benzo[a]anthracene	40—70
Chrysene	40—60
Benzo[e]pyrene	5—40
3-Methylfluoranthrene	40
2-Methylfluoranthrene	30
1,6-Methylchrysene	10
2,3-Methylchrysene	7
Dibenz[a,h]anthracene	40
Benzo[b]fluoranthrene	30
Benzo[j]fluoranthrene	60
Indeno[1,2,3-cd]pyrene	4
Dibenzo[a,h]pyrene	?
Dibenzo[a,j]pyrene	?
Benzo[c]phenanthrene	?
Dibenz[a,c]anthracene	?

Adapted from Shephard, R. J., *The Risks of Passive Smoking*, Oxford University Press, New York, 1982.

Nicotine levels can also vary significantly in tobacco smoke. Variables include the type of tobacco, growing conditions, pyrolysis temperature, and filter efficiency.[3] For example, Burley tobacco may contain up to 10% nicotine whereas Turkish tobacco usually has a maximum of 1%. Flue curing also reduces nicotine content. The temperature of pyrolysis will determine how much nicotine is converted to pyridine, myosine, bipyridyl, heterocyclic compounds, and aromatic hydrocarbons.

The major particulate matter of concern is tar which essentially is composed of various PAH compounds, most of which have known carcinogenic activity. As seen in Table 3, relatively large amounts of PAH can be found in cigarette smoke. Other data presented in Table 4 clearly show the dramatic differences in levels of toxic hydrocarbons that can be found in various tobacco smokes.

Acrolein is a pungent compound that is toxic to the tracheal ciliae and is a by-product of glycerin decomposition. Its level in tobacco smoke ranges from 45 to 140 mg per cigarette and is influenced by the nitrate, wax, polyphenol, and alkaloid content of the tobacco leaf.[3] Its concentration in smoke also increases with a decrease in cigarette length.

Table 4
TOXIC HYDROCARBONS (μg/100g OF TOBACCO) IN VARIOUS TOBACCO FORMS

	Smoke from		
Compound	**Cigarettes**	**Cigars**	**Pipe**
Anthracene	10.9	11.9	110.0
Pyrene	12.5	17.6	75.5
Acenaphthylene	5.0	1.6	29.1
3,4-Benzpyrene	0.9	3.4	8.5
Total	29.3	34.5	223.1

Adapted from Campbell, J. M. and Lindsey, A. J., *Br. J. Cancer*, 11, 192, 1957.

Various phenols that are common constituents of most all smoke include the compounds phenol, the three isomers of cresol, catechol, hydrofuranone, and resorcinol. They are toxic to ciliae and can contribute to carcinogenic changes to exposed cells. A wide variety of tobacco components, including glucose, sucrose, starch, cellulose, pectin, rutin, docosane, and tetracosane, can serve as precursors for the formation of phenols. The compound phenol can be produced in Oriental tobacco at levels ranging from 0.24 to 0.36 mg per cigarette, while in Burley tobacco the average level is only 0.04 mg per cigarette.[3] Cresol isomer levels are fairly constant among tobacco sources, averaging 0.07 to 0.1 mg per cigarette. Phenols in general can be rather effectively removed by cellulose acetate filters.[3]

The presence of hydrogen cyanide in tobacco smoke is well documented and is derived from the pyrolysis of proteins and amino acids. The concentration of hydrogen cyanide can increase threefold from first to last puff, and certain filters can become saturated with hydrogen cyanide before a cigarette is finished[14,15] thereby permitting inhalation of 0.1 to 0.4 mg per cigarette. The most effective filter for hydrogen cyanide appears to be the charcoal filter.[15] At a temperature around 1000°C there apparently is almost a total conversion of nitrogen compounds to hydrogen cyanide but yields decrease dramatically as the pyrolysis temperature falls below 800°C.[32]

Various oxides of nitrogen, including nitric oxide, nitrogen dioxide, nitrous oxide, and methyl nitrite, are present in tobacco smoke.[16,17] In fresh tobacco smoke, nitric oxide is the major component. This compound has a half-life of approximately 10 min at which time it is converted to nitrogen dioxide. Both forms of the gas are toxic and also have a synergistic influence in the formation of carcinogenic nitrosamines. Tobacco leaves contain up to 6% nitrate, which is the primary source of nitrogen oxides. Nitrate content in turn is influenced by the type and amount of fertilizer used to grow tobacco. Younger tobacco leaves contain less nitrate than older ones.[18] Cigarette smoke can have up to 0.6 mg per cigarette of nitric oxide and depending on the age of the smoke, a cigarette can have up to 0.01 mg of nitrogen dioxide.

Numerous nitrogen-containing compounds, including ammonia, primary amines such as methylamine, ethylamine, and butylamine, and secondary amines including dimethylamine and pyrrolidine, have been found in cigarette smoke.[13] The secondary amines in turn are of special concern since they can actively participate in the formation of nitrosamines.[19-21] The use of chemicals like maleic hydroazide to control the formation of suckers during tobacco plant growth and enzymatic activity during curing also enhance nitrosamine formation. Traces of aromatic amines such as aniline and naphthylamine can be present in tobacco smoke,[22] as well as certain mutagenic amino β-carbolines.[3] Various nitriles, including acetonitrile,

Table 5
PRODUCTION OF BENZO[A]PYRENE (B[A]P) FROM THE PYROLYSIS OF VARIOUS PRODUCTS

Starting material	μg B[a]P/100 g starting material
Agar	47
Hydroxyethylcellulose	34
Gum arabic	32
Sorbitol	13
Starch	7
Glycerol	6
Triethylene glycol	3

Adapted from Wakeham, H., *The Chemistry of Tobacco and Tobacco Smoke*, Schmeltz, I., Ed., Plenum Press, New York, 1972, 1.

proprionitrile, acrylonitrile, crotononitrile, and methacrylonitrile[3] can also be present, with the latter group of three compounds being formed from the pyrolysis of pyridine. Other nitrogen-containing alkaloids besides nicotine are present, including nornicotine, anabasine, anatabine, myosine, and cotinine.[3] Nitrogen-containing heterocyclic compounds include quinoline, indole, pyrrol, skatole, pyrazines, dibenzacridine, and dibenzcarbazole.[3]

A wide variety of oxygen-containing compounds are also present.[3] The vapor phase of tobacco smoke contains fairly large amounts of methanol[23] while numerous long-chained primary alcohols are present in the particulate phase. Apparently, sterols such as stigmasterol, β-sitosterol, compesterol, and cholesterol can appear in cigarette smoke in their unaltered form.[24] Important polyphenols present include scopoletin[25] and catechol,[26] the latter of which may serve as a tumor promoter and carcinogen. Many carbonyl compounds exist but acetaldehyde, formaldehyde, furfural, acetone, methacrolein, and 2,3-butanedione contribute to tobacco taste.[23] A standard complement of volatile and nonvolatile fatty acids are also present.[3]

Sulfur compounds have also been identified in cigarette smoke.[27] Some of the important ones include hydrogen sulfide, carbonyl sulfide, methyl thionitrile, carbon disulfide, sulfur dioxide, and thiophenes. Interestingly, the level of hydrogen sulfide found in cigarette smoke usually exceeds industrially permitted limits.

A full complement of hydrocarbons are present, including alkanes, alkenes, alkines, cyclic hydrocarbons, and aromatic hydrocarbons. Most of these compounds are derived from the waxy coating found in tobacco leaves, as well as cellulose, lignin, and pectins present.[28]

Trace elements that are commonly found in tobacco smoke include iron, antimony, zinc, and selenium. Other trace elements of potential concern include mercury from spray and paper, lead from sprays, and radium from phosphorus fertilizers.[3]

Other miscellaneous types of compounds that can unintentionally find their way into tobacco, and thus its smoke, include pesticides, insecticides, viruses, and aflatoxins.[29-31]

Because of the complexity of tobacco and its resulting smoke, in most situations it is extremely difficult to positively identify a precursor for certain compounds. For example, as shown in Table 5, the harmful compound benzo[a]pyrene can be demonstrated to be formed from the pyrolysis of numerous products[32] and thus, if one is attempting to decrease PAH content, one is not sure which precursor should be reduced. Similar data, as summarized in Table 6, have been reported for the formation of phenol.

Both large and small molecular weight compounds can be present in tobacco smoke, with the presence of certain large molecular weight compounds being somewhat difficult to

Table 6
PRODUCTION OF PHENOL FROM THE PYROLYSIS OF VARIOUS CARBOHYDRATES

Starting material	Conversion to phenol (%)
Sucrose	6.7
Glucose	5.0
Pectin	2.6
Cellulose	2.4
Rutin	0.8
Starch	0.2

Adapted from Hoegg, U. R., *Environ. Health Perspect.*, 2, 117, 1972.

explain. It has been postulated that these compounds can find their way into smoke because under thermal stress, tobacco cells may literally explode and thus eject nonvolatile solids into the smoke stream.[32] Therefore, compounds such as intact proteins, starch, pectin, and chlorophyll, which would not be expected to be present because of their large molecular weight, can be found in cigarette smoke.

V. MAINSTREAM VS. SIDESTREAM TOBACCO SMOKE

Since the active puffing of a cigarette dramatically influences smoke composition due to changes in the amount of oxygen present for pyrolysis, and pyrolysis temperature itself is dramatically altered, as compared to a resting cigarette, some researchers have divided smoke into mainstream and sidestream fractions. Maintream smoke can be defined as that smoke which is inhaled by the smoker and then exhaled into the air. During inhalation or puffing, pyrolysis temperature is at its highest point. Sidestream smoke in turn is defined as the smoke which is not inhaled, and normally represents around 95% of the smoke produced by a cigarette.[3] In addition, other forms of smoke have been identified including smolder smoke, defined as the smoke that emerges from the butt end of a cigarette between puffs. This portion becomes important since the filter is exposed to it even when the cigarette is not being puffed. Three other minor fractions include the glow stream, which is emitted by the glowing cone during puffing, the effusion stream, which escapes along the length of the paper during puffs, and the diffusion stream, which escapes through the paper between puffs.

The consumption of a typical cigarette usually involves ten 2-sec puffs, representing a total of 20 sec of puffing as mainstream smoke, and 500 sec of sidestream smoke. However, because cigarette combustion is faster during puffing, almost half of the tobacco is pyrolyzed in the 20-sec portion.[33]

Most of the constituents of mainstream smoke are retained by the human body. For example, if cigarette mainstream smoke is held in the mouth for 2 sec, retention of water-soluble compounds such as acetonitrile, acetaldehyde, and acetone is in the range of 48 to 78%, however, if the same is inhaled into the lungs, retention increases to 74 to 99%.[34,35] Also, exhaling through the nose results in more constituent retention than exhaling through the mouth.

The physical properties of mainstream and sidestream smokes are also different, with mainstream smoke containing on the average 5×10^9 particles per milliliter of smoke that have an average size of 0.2 μm.[36] As mentioned above, most of this particulate matter is retained by the lung, but that portion which is exhaled is larger in particle size due to

Table 7
RATIO OF CONCENTRATIONS OF VARIOUS CIGARETTE COMPONENTS BETWEEN SIDESTREAM AND MAINSTREAM SMOKE[21,32,33]

Compound	Sidestream/maintream ratio
Hydrogen cyanide	0.03
Acetic acid	0.42
Acetamide	0.68
Acetylene	0.80
Carbon monoxide	1.3—3.0
Benzene	1.9
Methyl chloride	2.1
3,4-Benzo[a]pyrene	2.1
Phenol	2.1
Acetonitrile	2.3
Nicotine	2.5
Toluene	3.5
Pyrene	3.6
Nitrous oxide	3.6
Carbon dioxide	3.8—10.0
Myosine	7.0
Pyrrole	7.1
3-Vinylpyridine	16.0
Naphthalene	16.0
Aniline	30.0
Ammonia	98.0

condensation. In contrast, the average particulate size of sidestream smoke is 0.15 μm and approximately 6.3×10^9 particles are produced for every second of cigarette-free burning.[36]

As would be expected, there are major chemical differences in mainstream and sidestream smoke fractions. Some of these more dramatic differences among compounds are summarized in Table 7. The question also arises as to the safety of nonintentional inhalation of sidestream smoke due to its environmental presence in room air space.

REFERENCES

1. **Baker, R. R.,** Temperature distribution inside a burning cigarette, *Nature,* 247, 405, 1974.
2. **Carter, W. L. and Hasegawa, I.,** Fixation of tobacco smoke aerosols for size distribution studies, *J. Colloid Interface Sci.,* 53, 134, 1975.
3. **Shephard, R. J.,** *The Risks of Passive Smoking,* Oxford University Press, New York, 1982.
4. **Tso, T. C. and Gori, G. B.,** Effect of tobacco characteristics on cigarette smoke composition, in *The Chemistry of Tobacco and Tobacco Smoke,* Schmeltz, I., Ed., Plenum Press, New York, 1972, 51.
5. **Osman, S. and Barson, J.,** Hydrocarbons of cigar smoke, *Tobacco,* 159, 30, 1964.
6. **Osman, S., Schmeltz, I., Higman, H. C., and Stedman, R. L.,** Volatile phenols of cigar smoke, *Tobacco,* 157, 30, 1963.
7. **Hoffmann, D., Rathkamp, G., and Wynder, E. L.,** Composition of the yields of several selected components in the smoke from different products, *J. Natl. Cancer Inst.,* 31, 627, 1963.
8. **Hoffmann, D. and Wynder, E. L.,** Smoke of cigarettes and little cigars: an analytical comparison, *Science,* 178, 1197, 1972.
9. **Armitage, A. K. and Turner, D. M.,** Absorption of nicotine in cigarette and cigar smoke through the oral mucosa, *Nature,* 226, 1231, 1970.

10. **Wakeham, H. R. R.,** Environmental carbon monoxide from cigarette smoke — a critique, *Prev. Med.*, 6, 526, 1977.
11. **Campbell, J. M. and Lindsey, A. J.,** Polycyclic hydrocarbons in cigar smoke, *Br. J. Cancer*, 11, 192, 1957.
12. **Johnson, W. R., Hale, R. W., Nedlock, J. W., Grubbs, H. J., and Powell, D. H.,** The distribution of products between mainstream and sidestream smoke, *Tobacco Sci.*, 17, 141, 1973.
13. **Johnson, W. R., Hale, R. W., Clough, S. C., and Chen, P. H.,** Chemistry of the conversion of nitrate nitrogen to smoke products, *Nature*, 243, 223, 1973.
14. **Williamson, J. T. and Allman, D. R.,** Effect of cigarette filter on the components of the vapor phase. Analysis of successive smoke puffs, *Beitr. Tabakforsch.*, 2, 263, 1964.
15. **Williamson, J. T., Graham, J. F., and Allman, D. R.,** The modification of cigarette smoke by filter tips, *Beitr. Tabakforsch.*, 3, 233, 1965.
16. **Norman, V. and Keith, C. H.,** Nitrogen oxides in tobacco smoke, *Nature*, 205, 915, 1965.
17. **Haagen-Smit, A. J., Brunelle, M. F., and Hara, J.,** Nitrogen content of smokes from different types of tobacco, *AMA Arch. Ind. Health*, 20, 399, 1959.
18. **Neurath, G. and Ehmke, H.,** Examination of nitrate contents of tobacco, *Beitr. Tabakforsch.*, 2, 333, 1964.
19. **Magee, P. N. and Barnes, J. M.,** Carcinogenic nitroso compounds, *Adv. Cancer Res.*, 10, 163, 1966.
20. **Fine, D. H., Rufeh, F., Lieb, D., and Rounbehler, D. P.,** Description of the thermal energy analyzer (TEA) for trace determination of volatile and non-volatile *N*-nitroso compounds, *Anal. Chem.*, 47, 1188, 1975.
21. **Brunnemann, K. D., Yu, L., and Hoffmann, D.,** Assessment of carcinogenic volatile *N*-nitrosamines in tobacco and in maintream and sidestream smoke from cigarettes, *Cancer Res.*, 37, 3218, 1977.
22. **Masuda, Y. and Hoffmann, D.,** Quantitative determinations of 1-naphthylamine and 2-naphthylamine in cigarette smoke, *Anal. Chem.*, 41, 650, 1969.
23. **Stedman, R. L.,** The chemical composition of tobacco and tobacco smoke, *Chem. Rev.*, 68, 153, 1968.
24. **Grunwald, C., Davis, D. L., and Bush, L. P.,** Cholesterol in cigarette smoke condensate, *J. Agric. Food Chem.*, 19, 138, 1971.
25. **Watanabe, R., McIlrath, W. J., Skok, J., Chroney, W., and Wender, S. H.,** Accumulation of scopoletin in boron-deficient tobacco leaves, *Arch. Biochem. Biophys.*, 94, 241, 1961.
26. **Carter, W. L. and Hasegawa, I.,** Fixation of tobacco smoke aerosols for size distribution studies, *J. Colloid Interface Sci.*, 53, 134, 1975.
27. **Horton, A. D. and Guerin, M. R.,** Quantitative determination of sulfur compounds in the gas phase of cigarette smoke, *J. Chromatogr.*, 90, 63, 1974.
28. **Gilbert, J. A. S. and Lindsey, A. J.,** The thermal decomposition of some tobacco constituents, *Br. J. Cancer*, 11, 398, 1957.
29. **Corbaz, R., Artho, A., Ceschini, P., Hausermann, M., and Plantefeve, J. C. I.,** Herbicide residues in tobacco leaves and their transfer into the smoke, *Beitr. Tabakforsch.*, 5, 80, 1969.
30. **Hengy, H. and Thirion, J.,** The determination of malathion residues in tobacco smoke condensate, *Beitr. Tabakforsch.*, 5, 175, 1970.
31. **Pattee, H. E.,** Production of aflatoxins by *Aspergillus flavus* cultured on flue-cured tobacco, *Appl. Microbiol.*, 18, 952, 1969.
32. **Wakeham, H.,** Recent trends in tobacco and tobacco smoke research, in *The Chemistry of Tobacco and Tobacco Smoke*, Schmeltz, I., Ed., Plenum Press, New York, 1972, 1.
33. **Hoegg, U. R.,** Cigarette smoke in closed spaces, *Environ. Health Perspect.*, 2, 117, 1972.
34. **Dalhamn, T., Edfors, M. L., and Rylander, R.,** Mouth absorption of various compounds in cigarette smoke, *Arch. Environ. Health*, 16, 831, 1968.
35. **Dalhamn, T., Edfors, M. L., and Rylander, R.,** Retention of cigarette smoke components in human lungs, *Arch. Environ. Health*, 17, 746, 1968.
36. **Keith, C. H. and Derrick, J. C.,** Measurement of the particle size distribution and concentration of cigarette smoke by the "confuge", *J. Colloid Sci.*, 15, 340, 1960.

INDEX

D

E

F

G

H

Q

R

S